Advancements in Gel Science

Advancements in Gel Science—A Special Issue in Memory of Toyoichi Tanaka

Special Issue Editor

Masayuki Tokita

MDPI • Basel • Beijing • Wuhan • Barcelona • Belgrade

MDPI

Special Issue Editor
Masayuki Tokita
Kyushu University
Japan

Editorial Office
MDPI
St. Alban-Anlage 66
4052 Basel, Switzerland

This is a reprint of articles from the Special Issue published online in the open access journal *Gels* (ISSN 2310-2861) from 2018 to 2019 (available at: https://www.mdpi.com/journal/gels/special_issues/gel_science)

For citation purposes, cite each article independently as indicated on the article page online and as indicated below:

LastName, A.A.; LastName, B.B.; LastName, C.C. Article Title. *Journal Name* **Year**, *Article Number*, Page Range.

ISBN 978-3-03921-343-6 (Pbk)
ISBN 978-3-03921-344-3 (PDF)

Cover image courtesy of Masayuki Tokita.

Contents

About the Special Issue Editor

Masayuki Tokita received his Ph.D. from Hokkaido University in 1983. He then became an assistant of Hokkaido University, holding this position from 1983 to 1991. During this interval, he belonged to Toyoichi Tanaka's laboratory at the Massachusetts Institute of Technology as a visiting scientist from 1989 to 1991. He moved to Mie University in 1992 as an associate professor. In 2002, he moved to Kyushu University as a professor, and became a professor emeritus of Kyushu University in 2018.

gels

MDPI

Editorial

Advancements in Gel Science—A Special Issue in the Memory of Toyoichi Tanaka

Masayuki Tokita

Professor Emeritus, Kyushu University, Fukuoka, Japan; northbear3.14@gmail.com

Received: 20 June 2019; Accepted: 24 June 2019; Published: 26 June 2019

It is a great pleasure for us to present a collection of recent papers that were submitted to the special issue of Gels, Advancements in Gel Science—A Special Issue in Memory of Toyoichi Tanaka. The science of gel attracts much attention from many scientists who work in the fields of physics, chemistry, and biology. The famous chemist, P. J. Flory at Stanford University, has built the fundamentals of the gel science. His brilliant achievements are published in many journals and, finally, knowledge about the science of not only gel, but also the physical chemistry of polymers is summarized in his "Bible", Principles of Polymer Chemistry, which was published in 1953 by Cornell University Press.

In 1977, the physicist Toyoich Tanaka at Massachusetts Institute of Technology achieved a great breakthrough in gel science, with the discovery of the critical phenomena and the volume phase transition of gels. After the discovery, Toyo, which was his nickname in the lab, cultivated the new era of gel science. His science is based on experimental study, though he also had deep insight into theoretical physics as well as biological physics. Therefore, the papers he published are excellent guides for researches who are working in the area of physics, chemistry, and biology, even today. Many scientists expected further expansion of Toyo's gel science, however Toyo suddenly passed away on 20 May 2000 at the age of 54. His papers that are related to significant discoveries were selected from his publications, which expanded most areas of science, and were published in 2002 by Tokyo University Press as a collection of Toyo's posthumous works, named From Gels to Life.

The year 2017 was the 40th anniversary for the discovery of the volume phase transition of gel. It is, therefore, a timely opportunity to summarize the recent advancements of gel science from this year. As a result, we, MDPI, planned to publish a special issue of Gels. In this special issue, fortunately, 14 papers on the recent advancements of gel science in the areas of physics, chemistry, and biological science are summarized. I believe that all papers published here are of interest scientifically and that they have a significant impact on each area of science. We thank all scientists who cooperated with our plan and submitted their recent research work on gels to this special issue. We hope that this special issue becomes a milestone for gel science for the next 10 years.

Review

Single Micrometer-Sized Gels: Unique Mechanics and Characters for Applications

Miho Yanagisawa [1,*], Chiho Watanabe [1] and Kei Fujiwara [2]

[1] Department of Applied Physics, Tokyo University of Agriculture and Technology, Naka-cho 2-24-16,
    Koganei, Tokyo 184-8588, Japan; cwatanabe@m2.tuat.ac.jp
[2] Department of Biosciences and Informatics, Faculty of Science and Technology, Keio University,
    3-14-1 Hiyoshi, Yokohama 223-8522, Japan; fujiwara@bio.keio.ac.jp
* Correspondence: myanagi@cc.tuat.ac.jp

Received: 27 February 2018; Accepted: 26 March 2018; Published: 28 March 2018

Abstract: Microgels—small gels of submicron to micron size—are widely used in food, cosmetics and biomedical applications because of their biocompatibility and/or fast response to external environments. However, the properties of "single" microgels have not been characterized due to limitations in preparation technologies and measurement methods for single microgels with sizes in the multi-micrometer range. The synthesis of multiple shapes of single microgels and their characterization are important for further functionalization and application of gel-based materials. In this review, we explain the recent advancements in microgel fabrication and characterization methods for single microgels. The first topic discussed includes the self-assembly methods for single microgel fabrication using physical phenomena such as phase separation, interfacial wetting and buckling instability. The second topic deals with methods for analyzing the mechanics of single microgels and the differences between their mechanical characteristics and those of bulk gels. The recent progress in the fabrication and characterization of single microgels will bring important insights to the design and functionalization of gel-based materials.

Keywords: janus particle; anisotropic shape; phase separation; wetting; micrometric confinement; micropipette aspiration

---

1. Introduction

Polymer gels are viscoelastic materials that can store large amounts of liquid in their polymer networks. The liquid, which constitutes the majority of the polymer gel, contributes to their soft and hydrophilic properties. Because of these characteristics, polymer gels have become ubiquitous and indispensable. For example, the biocompatibility and biodegradability of hydrogels are beneficial for biomedical applications [1,2]; the controllable nature of their refractive index is applied in soft contact lenses [3] and optical lenses [4]; the functional hydrophilic material entrapped in gels is used in biomedical cosmetics and food materials [5]. Furthermore, a gel composed of a hydrophobic polymer has been featured as an oil absorbing agent for environmental recovery [6]. The growth in the versatile usage of polymer gels has prompted the necessity for improving polymer gels.

With progress in the functionalization of polymer gels, the sizes of gels have recently featured in a novel method of regulating the characteristics of gels. On account of their larger surface to volume ratios compared to bulk gels, smaller gels exhibit higher permeabilities, promoting efficient substance transport between gels and their environments [7,8]. This feature is prominent in small gels of submicron to micron size, that is, microgels. The fast response of microgels to external environments is applied in repairing damaged cell tissues [9–13] and the smooth release of the drugs trapped inside them [14].

The question that needs to be answered is whether a single microgel has the same characteristics as a bulk gel. Although the higher permeabilities of microgels compared to bulk gels suggests a negative answer, the elucidation of microgel properties should contribute greatly to the further functionalization of gels.

Single microgels have not been characterized due to the limitations in measuring methods, unlike bulk gels and solutions containing microgels [15,16]. Recent studies have shown that single microgels exhibit unique mechanics and characters that are beneficial for numerous applications (Figure 1). For example, single microgels can generate functions according to their shapes and stabilize artificial cells like cytoskeletons in cells [17,18]. In this review, we introduce the recent advances in the preparation, characterization and application of single micrometer-sized gels using water-in-oil emulsions. Although research on nanometer-sized gels [19] is progressing and their characteristics have been analyzed using super resolution microscopy, scattering techniques and so forth. [20,21], such gels are not the focus of this review.

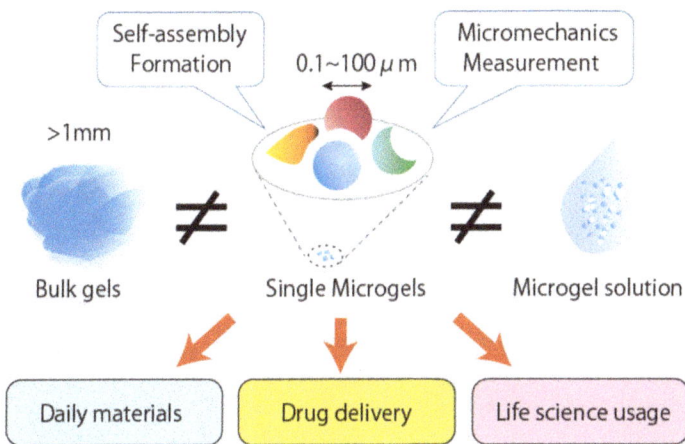

Figure 1. Schematic illustration of nonspherical single microgels exhibiting unique properties and applications.

2. Morphology of Single Microgels

A major strategy for preparing microgels is the gelation of polymers inside microsized emulsions (emulsion polymerization). In general, microgels synthesized by gelation inside emulsions have spherical or core-shell shapes. Although the sphere-based shape is a common morphology for microgels, the synthesis of non-spherical particles has gained attention due to their potential for multiple applications, for example, as building blocks for complex assemblies [22–24], drug carriers and functional coatings [25]. Gelation inside a microsized mold is a technique to prepare non-spherical microgels. However, the preparation of molds is a laborious process. A recent study used physical phenomena such as polymer phase separation and droplet surface interaction as alternative morphology-controlling method for synthesizing non-spherical microgels [26]. This method is based on spontaneous spatiotemporal organization and therefore is expected to pave the way for forming microgels with desired morphologies without complicated molecular synthesis or expensive equipment.

2.1. Morphology Control through Phase Separation and Gelation of Polymers inside Microdroplets

In emulsion polymerization, the mold for gelation is usually spherical because it is the most stable shape for lipid droplets with minimum surface area. The gelation of the polymers confined in spherical droplets results in spherical microgels. The use of liquid-liquid phase separation polymers helps to obtain microgels of various shapes without designing molds. Here, we introduce a method for obtaining non-spherical shaped microgels using microdroplets (water-in-oil emulsion covered with a surfactant layer such as lipid).

In the bulk (not in microdroplets), the phase separation of binary polymer solutions causes the appearance of spherical domains that minimizes the interfacial energy before the separation into two phases (Figure 2(ai)) [27]. However, spherical gels are not stable for a few reasons. The gelation space is smaller than the domain size. Since phase separation is trapped by fast gelation during the separation process, the rate of phase separation relative to gelation and/or the timing of gelation are also factors that control the shapes of microgels. The space of gelation affects the interfacial tension between the coexisting phases to alter the affinity (wettability) of the polymer on the surface of the confining space. Therefore, the phase separation and gelation of polymers inside microdroplets produce microgels of different shapes [28]. The degradation of affinity (de-wetting transition) enables the shape deformation of the phase separating droplets and their separation into two isolated droplets [29].

In this section, we explain in detail the method for producing nonspherical microgels using phase separation from the following two viewpoints: (i) gelation and wetting of polymers on the microdroplet surface; and (ii) volume fraction of coexisting phases and size of confinement space. In addition to the phase separation method, it has been shown that nonspherical microgels are formed using microfluidic devices. In this review, we do not explain the details; alternatively, we introduce review articles [30–33].

2.1.1. Gelation and Wetting of Polymers on the Microdroplet Surface

Non-spherical microgels have been fabricated by the gelation of gelling-polymer-rich domains upon phase separation and wetting in microdroplets (Figure 2b) [34]. The domain morphology after the completion of liquid-liquid phase separation is determined by the wetting of the phases coexisting at the microdroplet interface. The degree of wetting is determined by the contact angle, which essentially determines the balance between the interfacial tensions existing among the coexisting phases. In the case of phase separating solutions of binary polymers inside microdroplets, the wettability is determined by the interfacial tensions between the coexisting polymers and also between the polymers and interface of the confined microdroplets. When gelatin and a poly(ethylene glycol) (PEG) solution are confined in lipid microdroplets, the wettability of gelatin depends on the type of lipid covering the droplets (Figure 2(aii)). If the two coexisting polymers have different affinities for the microdroplet interface, the shapes of the microgels will drastically change according to the characters of the interfaces, which are controlled by changing the surfactants at the interface.

However, controlling the wetting property of a gelling polymer is not easy since the surface tension between coexisting phases changes before and after gelation. For example, Ma et al. reported that the phase separation pattern changes before and after UV-induced gelation of poly-(ethylene glycol) diacrylate (PEGDA) (Figure 3A) [35]. This suggests that the selective polymerization of coexisting phases alters the interfacial tension between coexisting phases and accordingly changes the wettability and contact angle. In addition, gelation increases the elasticity of gelling-polymer-rich domains. Thus, various complex patterns are formed by controlling the wetting property, especially for a system whose phase separation and gelation simultaneously progress with temperature change [36,37].

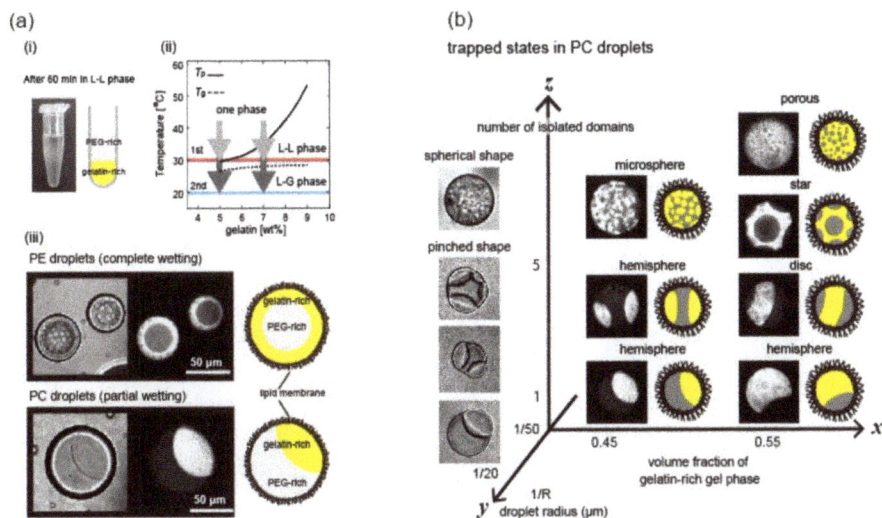

Figure 2. Morphology control using phase separation and gelation of gelatin inside microdroplets. (a) (i) An example of phase-separated solution of poly-(ethylene glycol) (PEG)1.7 wt % and gelatin 5.0 wt % in tube; (ii) Schematic phase diagram of PEG/gelatin blends containing the PEG 1.7 wt % and various concentrations of gelatin. The black solid and dashed lines indicate the phase separation point, *T*p and the gelation point, *T*g, respectively; (iii) Phase separation of PEG and gelatin 5.0 wt % blend in phosphatidylethanolamine (PE) and phosphatidylcholine (PC) droplets after 60 min incubation. Droplets are shown in (from left to right) differential interference contrast (DIC) images, fluorescent images of the gelatin-rich gel phase and schematic illustrations. (b) Variously shaped microgels in PC droplets in response to changes in volume fraction of gelatin-rich phase, droplet size and number of isolated domains. In the *x*–*z* plane, large droplets with *R* ~50 μm are shown in fluorescent images and schematic illustrations, where gelatin-rich phase is shown in white and yellow, respectively. In the *y*–*z* plane, small droplets with *R* ~20 μm are shown as DIC images. (M. Yanagisawa et al., Copyright 2014, National Academy of Sciences).

2.1.2. Volume Fraction of Two Coexisting Phases and Size of the Confining Microdroplet

At a constant contact angle, changes in volume fractions can vary the shapes of single microgels (Figures 2b and 3B). In the case of partial wetting, concave single microgels are formed. The crenated shape of a microgel depends on the volume fraction of the gelling-polymer rich phase [34,35]. Although such concave microgels can be formed by adding smaller particles as a mold, the advantage of the phase separation method is that it does not require mold particles [23]. The size of the microdroplet affects the microgel shape, since the balance between bending energy and interfacial energy depends on the microdroplet size (Figures 2b and 3C) [35,36].

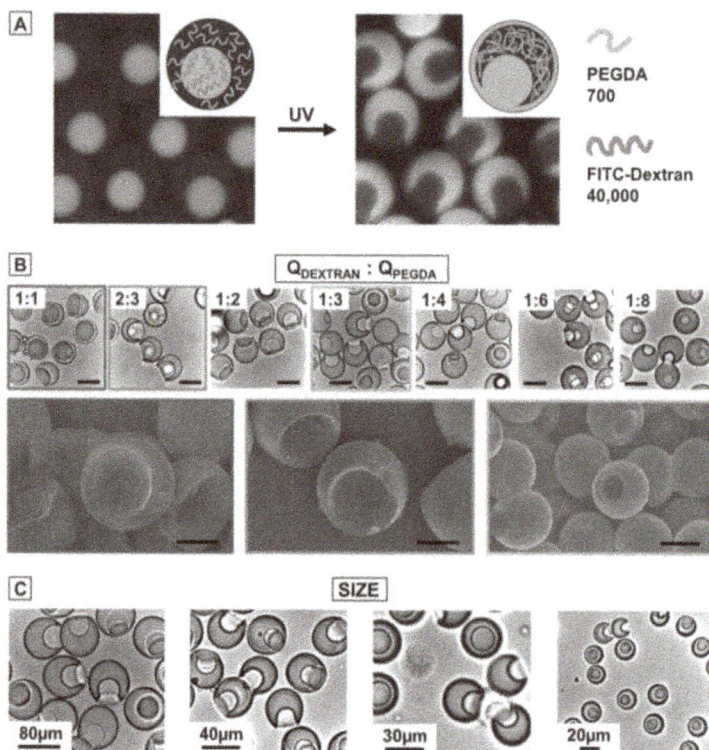

Figure 3. Selective polymerization of dextran/ poly-(ethylene glycol) diacrylate (PEGDA) core–shell droplets. (**A**) Fluorescence images of phase-separated aqueous two-phase system (ATPS) droplets consisting of a fluorescently labeled dextran core and a non-labeled PEGDA shell. During polymerization, dextran migrates into the PEGDA shell. As the core is depleted of dextran, the resulting microgel particle contains a socket. (**B**) Bright-field microscopy images of gel microparticles with a socket. The socket size is determined by the flow rate ratio of dextran and PEGDA (upper row). Scanning electron micrographs of selected samples (lower row). The scale bars denote 80 μm in the upper row and 30 μm in the lower row. (**C**) Bright-field microscopy images of particles with different sizes. While the overall flow rate of the ATPS mixture and the fluorinated oil is kept constant, the total particle size is determined by the flow rate ratio of the ATPS mixture and the fluorinated oil phase as well as the height of the second microfluidic nozzle, which is either 75 or 20 μm. (S. Ma et al., Copyright 2012 WILEY-VCH Verlag GmbH & Co. KGaA, Weinheim).

2.2. Morphology Control by Buckling Instability

Buckling is another useful phenomenon for forming non-spherical microgels (Figure 4). Buckling instability takes place when a layer is subjected to contraction [38]. One strategy for inducing contraction is to form a heterogeneous surface gel network by adding agents that change the cross-linking gelling rate [33]. Microgels with heterogeneous crosslinks of polymer networks swell heterogeneously, leading to buckle formation. Another method is to use specially designed monomers with different cross-linking moieties [22,23,25].

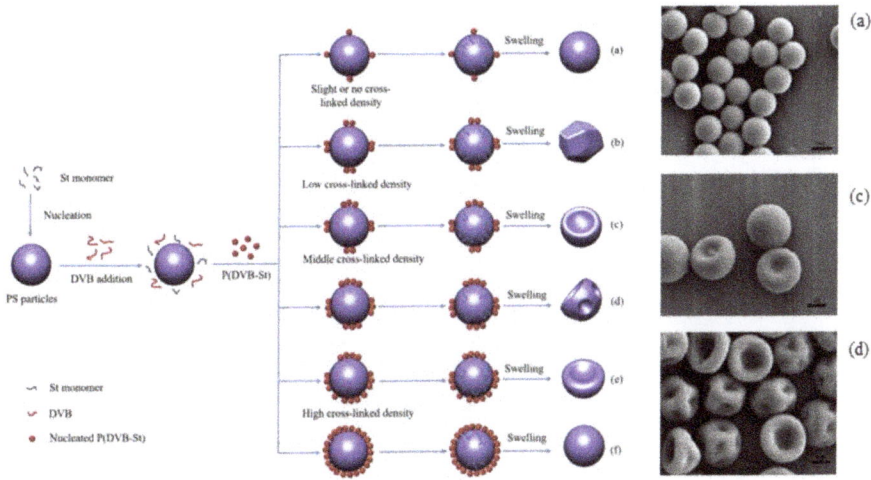

Figure 4. Nonspherical single microgels prepared using buckling instability. (**left**) Possible formation mechanism of the buckling of microgels with various surface morphologies; (**right**) Scanning electron microscopy (SEM) micrographs of microgels synthesized by varying the divinylbenzene (DVB) concentration based on total styrene (St) monomer mass, (**a**) 0 wt %, (**c**) 0.5 wt % and (**d**) 2.0 wt %. The scale bars are 1 μm. (H. Shen et al., Copyright 2017 ELSEVIER).

Using two consecutive polymerizations, it forms soft core-hard shell structures. Consequently, mismatches between the inner and outer mechanical properties induce buckling of the surface shell. Buckled microgels can be useful for colloidal building blocks, along with smaller spherical microgels [24].

3. Micromechanics Measurement of Single Microgels

Microgels are soft and deformable microparticles. As mentioned above, it is important to elucidate the mechanical properties of single microgels rather than their solution for application purposes. However, in contrast to the accumulation of the macroscopic mechanical properties of microgel solutions, the microscopic mechanical properties of single microgels were ambiguous. Experimental difficulty was the obstacle in analyzing the microscopic mechanical properties of a single microgel. To solve the problem, methods involving atomic force microscopy (AFM) [39–41] and microcapillary methods for analyzing the deformation during suction [42–46] and compression [42,47,48] have recently been employed (Figure 5).

In the case of AFM measurement, two technical difficulties remain: placing soft particles on a flat substrate and avoiding the deformation of the particle during pressing by AFM tips for measurement (Figure 5a). Recently, Mohapatra et al. reported that the induced Young's modulus of an air-dried microgel obtained using AFM was different from that obtained using Brillouin light scattering (BLA) (Table 1) [40]. Such AFM measurements become more difficult in the case of microgels floating in a liquid. In contrast, when using microcapillaries, it is easy to measure the microgels floating in a liquid, since aspirating or trapping a single microgel in the capillary can be used to measure the elasticity (Figure 5b,c) [42]. In the following sections, we introduce the microscopic mechanical properties of single microgels, in particular their elasticity as revealed by the microcapillary method.

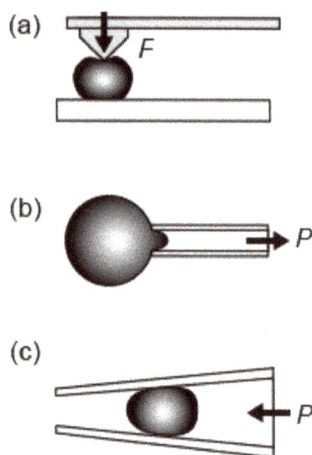

Figure 5. Schematic illustrations of methods for measuring the elastic properties of single microgels. (a) Atomic force microscopy (AFM) and (b,c) microcapillary. The single microgel is (b) aspirated inside the microcapillary and (c) deformed inside the microcapillary. The F and P represent the applied force by AFM tip and applied pressure by microcapillary, respectively.

3.1. Micromechanics of Single Microgels

In the emulsion polymerization method, the physicochemical environments around the monomers of the gels near the surface are different from that in other areas. Therefore, the balance between the rate of surface polymerization and diffusion of the polymerization field across the droplet is considered a factor determining its mechanical properties. In fact, the balance has been reported to change the shell thicknesses of core-shell microgels [2,49], which is an important parameter determining their mechanistic properties. For instance, the mechanical stability of a single microgel increases as shell thickness increases [50] and the elasticity varies depending on the size ratio of the core-to-shell thickness [48,49]. Although the correlation between the mechanical properties and structure of microgel is not yet clear, these reports demonstrate that the kinetics of gelation is important in determining the mechanical properties of single microgels.

In general, microgels with high surface areas/volume ratios exhibit immediate responses to changes in external environments and alter their structures and mechanical properties to fit their environments. In fact, the elasticity of the poly(n-isopropylacrylamide) (PNIPAM) microgel has been reported to change greatly depending on the surrounding environment (temperature, solvent, etc.) [51]. In addition, it was reported that the Young's modulus of PNIPAM microgel is larger than that of the corresponding bulk gel (Figure 6) [47]. Furthermore, Young's moduli of polystyrene-*co*-poly(n-isopropylacrylamide) (pS-*co*-NIPAM) microgels are sensitive to the cross-link concentration, unlike the bulk modulus (Table 1) [40].

Recently, our group measured the elasticities of gelatin single microgels prepared inside microdroplets covered with lipid layers by micropipette aspiration [46]. We found that the elasticities of small microgels with radius less than 50 µm are higher than those of the corresponding bulk gels. In this study, the structural changes in gelatin molecules induced by the lipid membranes covering the microdroplets were pointed out as a factor for the increase in elasticity.

In addition, in non-spherical microgels (Figures 2 and 3), only a part of the microgel is in contact with the surfactant film, such as a lipid layer covering the droplet. Therefore, the shapes and surface dynamics of non-spherical microgels are expected to be anisotropic. Since microcapillary aspiration (Figure 5b) can measure the local elasticity of the microgel, the correlation between shape and local elasticity will be revealed in the near future.

As described above, the characteristics of lipid membranes may change the surface gel structure and the mechanical properties of the gels due to lipid membrane-polymer interactions. These effects should be the case for biological single microgels like cytoskeletons underneath cellular membranes. Thus, elucidation of the unique characters of single microgels and their mechanisms should contribute to understanding the nature of living cells.

Table 1. Comparison of Young modulus determined for the air-dried polystyrene-co-poly(*n*-isopropylacrylamide) (pS-*co*-NIPAM) microgel series with different cross-link concentration (represented by numbers following the name) using Brillouin light scattering (BLA) and AFM. (H. Mohapatra et al., Copyright 2017 Royal Society of Chemistry).

Microgel Sample	Bulk Modulus, K (GPa)	BLS Young's Modulus, E (GPa)	AFM Young's Modulus, E (GPa)
pS-*co*-NIPAM-1	4.92 ± 0.2	3.59 ± 0.14	0.24 ± 0.03
pS-*co*-NIPAM-2	4.86 ± 0.19	3.70 ± 0.15	0.54 ± 0.03
pS-*co*-NIPAM-3	4.86 ± 0.19	4.07 ± 0.16	0.61 ± 0.05
pS-*co*-NIPAM-4	4.92 ± 0.2	4.25 ± 0.16	0.91 ± 0.06
pS-*co*-NIPAM-5	5.02 ± 0.2	4.65 ± 0.19	0.99 ± 0.03

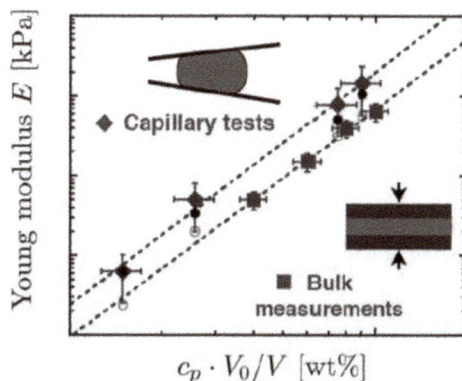

Figure 6. Young's modulus E from compressive mechanical tests on bulk polyacrylamide gels (squares) and for microgel particles (diamonds) as derived from the compressive modulus K (solid circles) and the shear modulus G (open circles). To account for the swelling of the particles in water, the polymer concentration cp is corrected by the volume ratio $V0/V$ ($V0$ is the initial volume of the particles as they are formed in oil and V is the equilibrium volume of the collected particles in water). (H. M. Wyss et al., Copyright 2010 Royal Society of Chemistry).

3.2. Structural Differences between Single Microgels and Bulk Gels

Why does a single microgel show gel elasticity different from that of the bulk gel? This is because gel elasticity is determined by the polymer network structure constituting the gel and the difference in gel elasticity indicates that the micro or nanostructures of microgels are different from those of bulk gels.

In the case of emulsion polymerization, gelation starts from the droplet surface of gelling polymers. A balance between the gelation and penetration rates of the cross-linking agent for gelation into the interior possibly renders the network structure of the microgels heterogeneous. This heterogeneous microgel structure could be the reason for the buckling of microgels (Figure 4) [40]. If this is the case, the surfactant film covering the droplet surface changes the interaction between the surface and the gelling polymer encapsulated inside. This effect will also result in a different microgel structure compared to the bulk gel. We recently found that the microgel elasticity is higher than that of the bulk gel as a result of changing its secondary structure by synthesizing a gelatin gel in lipid-film-covered microdroplets [46].

Biopolymers such as gelatin are known to correlate with lipids and similar materials in cell membranes [52]. Therefore, this progress indicates that it is possible to control the structures and elasticities of microgels via the correlation between the surfactant membrane covering the microdroplet surface and the gelling polymer. Since reducing the size of the microgel increases the ratio of surface area to internal volume, elastic control of microgels through size is also expected.

4. Summary and Perspective

Without complicating the molecular design of gelling polymers, it is now possible to change the shape of single microgels by utilizing phase separation between the non-gelling polymer and the gelling polymer inside microdroplets. By developing new methods to measure the elasticity of a single microgel, the unique mechanical properties of microgels (different from those of bulk gels) have been revealed. Further elucidation of the structure of the microgel and the process of gelation and functionalization utilizing the shape and mechanical properties of the microgel are expected in the near future.

Acknowledgments: This work was supported by JSPS KAKENHI (No. 15H05463, No. 15KT0081, No. 16H00796 and No. 16H01443).

Conflicts of Interest: The authors declare no conflict of interest.

References

1. Saunders, B.R.; Vincent, B. Microgel particles as model colloids: Theory, properties and applications. *Adv. Colloid Interface Sci.* **1999**, *80*, 1–25. [CrossRef]
2. Fernandez-Nieves, A.; Wyss, H.; Mattsson, J.; Weitz, D.A. *Microgel Suspensions: Fundamentals and Applications*; John Wiley & Sons: Hoboken, NJ, USA, 2011.
3. Calo, E.; Khutoryanskiy, V.V. Biomedical applications of hydrogels: A review of patents and commercial products. *Eur. Polym. J.* **2015**, *65*, 252–267. [CrossRef]
4. Vidyasagar, A.; Handore, K.; Sureshan, K.M. Soft Optical Devices from Self-Healing Gels Formed by Oil and Sugar-Based Organogelators. *Angew. Chem. Int. Ed.* **2011**, *50*, 8021–8024. [CrossRef] [PubMed]
5. McClements, D.J. Designing biopolymer microgels to encapsulate, protect and deliver bioactive components: Physicochemical aspects. *Adv. Colloid Interface Sci.* **2017**, *240*, 31–59. [CrossRef]
6. Hayase, G.; Kanamori, K.; Fukuchi, M.; Kaji, H.; Nakanishi, K. Facile synthesis of marshmallow-like macroporous gels usable under harsh conditions for the separation of oil and water. *Angew. Chem. Int. Ed.* **2013**, *52*, 1986–1989. [CrossRef] [PubMed]
7. Cai, W.S.; Gupta, R.B. Fast-responding bulk hydrogels with microstructure. *J. Appl. Polym. Sci.* **2002**, *83*, 169–178. [CrossRef]
8. Shimanovich, U.; Efimov, I.; Mason, T.O.; Flagmeier, P.; Buell, A.K.; Gedanken, A.; Linse, S.; Åkerfeldt, K.S.; Dobson, C.M.; Weitz, D.A.; et al. Protein microgels from amyloid fibril networks. *ACS Nano* **2015**, *9*, 43–51. [CrossRef] [PubMed]
9. Freemont, T.J.; Saunders, B.R. pH-Responsive microgel dispersions for repairing damaged load-bearing soft tissue. *Soft Matter* **2008**, *4*, 919–924. [CrossRef]
10. Velasco, D.; Tumarkin, E.; Kumacheva, E. Microfluidic encapsulation of cells in polymer microgels. *Small* **2012**, *8*, 1633–1642. [CrossRef] [PubMed]
11. Morimoto, Y.; Takeuchi, S. Three-dimensional cell culture based on microfluidic techniques to mimic living tissues. *Biomater. Sci.* **2013**, *1*, 257–264. [CrossRef]
12. Thomas, D.; Fontana, G.; Chen, X.; Sanz-Nogués, C.; Zeugolis, D.I.; Dockery, P.; O'Brien, T.; Pandit, A. A shape-controlled tuneable microgel platform to modulate angiogenic paracrine responses in stem cells. *Biomaterials* **2014**, *35*, 8757–8766. [CrossRef] [PubMed]

13. Jiang, W.; Li, M.; Chen, Z.; Leong, K.W. Cell-laden microfluidic microgels for tissue regeneration. *Lab Chip* **2016**, *16*, 4482–4506. [CrossRef] [PubMed]

14. Li, Y.; Meng, H.; Liu, Y.; Narkar, A.; Lee, B.P. Gelatin Microgel Incorporated Poly(ethylene glycol)-Based Bioadhesive with Enhanced Adhesive Property and Bioactivity. *ACS Appl. Mater. Interfaces* **2016**, *8*, 11980–11989. [CrossRef] [PubMed]

15. Suzawa, E.; Kaneda, I. Rheological properties of agar microgel suspensions prepared using water-in-oil emulsions. *J. Biorheol.* **2010**, *24*, 70–76. [CrossRef]

16. Lyon, L.A.; Fernandez-Nieves, A. The polymer/colloid duality of microgel suspensions. *Annu. Rev. Phys. Chem.* **2012**, *63*, 25–43. [CrossRef] [PubMed]

17. Kurokawa, C.; Fujiwara, K.; Morita, M.; Kawamata, I.; Kawagishi, Y.; Sakai, A.; Murayama, Y.; Shin-ichiro, M.N.; Murata, S.; Takinoue, M.; et al. DNA cytoskeleton for stabilizing artificial cells. *Proc. Natl. Acad. Sci. USA* **2017**, *114*, 7228–7233. [CrossRef] [PubMed]

18. Kumar, R.K.; Yu, X.; Patil, A.J.; Li, M.; Mann, S. Cytoskeletal-like supramolecular assembly and nanoparticle-based motors in a model protocell. *Angew. Chem. Int. Ed. Engl.* **2011**, *50*, 9343–9347. [CrossRef] [PubMed]

19. Plamper, F.A.; Richtering, W. Functional Microgels and Microgel Systems. *Acc. Chem. Res.* **2017**, *50*, 131–140. [CrossRef] [PubMed]

20. Di Lorenzo, F.; Seiffert, S. Nanostructural heterogeneity in polymer networks and gels. *Polym. Chem.* **2015**, *6*, 5515–5528. [CrossRef]

21. Wöll, D.; Flors, C. Super-resolution Fluorescence Imaging for Materials Science. *Small Methods* **2017**, *1*, 1700191. [CrossRef]

22. Sacanna, S.; Pine, D.J. Shape-anisotropic colloids: Building blocks for complex assemblies. *Curr. Opin. Colloid Interface Sci.* **2011**, *16*, 96–105. [CrossRef]

23. Sacanna, S.; Irvine, W.T.M.; Chaikin, P.M.; Pine, D.J. Lock and key colloids. *Nature* **2010**, *464*, 575–578. [CrossRef] [PubMed]

24. Sacanna, S.; Korpics, M.; Rodriguez, K.; Colón-Meléndez, L.; Kim, S.H.; Pine, D.J.; Yi, G.R. Shaping colloids for self-assembly. *Nat. Commun.* **2013**, *4*, 1688. [CrossRef] [PubMed]

25. Shen, H.; Du, X.; Ren, X.; Xie, Y.; Sheng, X.; Zhang, X. Morphology control of anisotropic nonspherical functional polymeric particles by one-pot dispersion polymerization. *React. Funct. Polym.* **2017**, *112*, 53–59. [CrossRef]

26. Glotzer, S.C.; Solomon, M.J. Anisotropy of building blocks and their assembly into complex structures. *Nat. Mater.* **2007**, *6*, 557–562. [CrossRef] [PubMed]

27. Yanagisawa, M.; Yamashita, Y.; Mukai, S.-A.; Annaka, M.; Tokita, M. Phase separation in binary polymer solution: Gelatin/poly (ethylene glycol) system. *J. Mol. Liq.* **2014**, *200*, 2–6. [CrossRef]

28. Kanamori, K.; Yonezawa, H.; Nakanishi, K.; Hirao, K.; Jinnai, H. Structural formation of hybrid siloxane-based polymer monolith in confined spaces. *J. Sep. Sci.* **2004**, *27*, 874–886. [CrossRef] [PubMed]

29. Choi, C.H.; Weitz, D.A.; Lee, C.S. One step formation of controllable complex emulsions: From functional particles to simultaneous encapsulation of hydrophilic and hydrophobic agents into desired position. *Adv. Mater.* **2013**, *25*, 2536–2541. [CrossRef] [PubMed]

30. Shum, H.C.; Abate, A.R.; Lee, D.; Studart, A.R.; Wang, B.; Chen, C.H.; Thiele, J.; Shah, R.K.; Krummel, A.; Weitz, D.A. Droplet microfluidics for fabrication of non-spherical particles. *Macromol. Rapid Commun.* **2010**, *31*, 108–118. [CrossRef] [PubMed]

31. Walther, A.; Muller, A.H.E. Janus Particles: Synthesis, Self-Assembly, Physical Properties and Applications. *Chem. Rev.* **2013**, *113*, 5194–5261. [CrossRef] [PubMed]

32. Shang, L.; Cheng, Y.; Zhao, Y. Emerging Droplet Microfluidics. *Chem. Rev.* **2017**, *117*, 7964–8040. [CrossRef] [PubMed]

33. Ma, J.Y.; Wang, Y.C.; Liu, J. Biomaterials Meet Microfluidics: From Synthesis Technologies to Biological Applications. *Micromachines* **2017**, *8*, 255. [CrossRef]

34. Hu, Y.; Wang, S.; Abbaspourrad, A.; Ardekani, A.M. Fabrication of shape controllable Janus alginate/pNIPAAm microgels via microfluidics technique and off-chip ionic cross-linking. *Langmuir* **2015**, *31*, 1885–1891. [CrossRef] [PubMed]

35. Ma, S.; Thiele, J.; Liu, X.; Bai, Y.; Abell, C.; Huck, W.T. Fabrication of microgel particles with complex shape via selective polymerization of aqueous two-phase systems. *Small* **2012**, *8*, 2356–2360. [CrossRef] [PubMed]

36. Yanagisawa, M.; Nigorikawa, S.; Sakaue, T.; Fujiwara, K.; Tokita, M. Multiple patterns of polymer gels in microspheres due to the interplay among phase separation, wetting and gelation. *Proc. Natl. Acad. Sci. USA* **2014**, *111*, 15894–15899. [CrossRef] [PubMed]
37. Zarzar, L.D.; Sresht, V.; Sletten, E.M.; Kalow, J.A.; Blankschtein, D.; Swager, T.M. Dynamically reconfigurable complex emulsions via tunable interfacial tensions. *Nature* **2015**, *518*, 520–524. [CrossRef] [PubMed]
38. Mora, T.; Boudaoud, A. Buckling of swelling gels. *Eur. Phys. J. E Soft Matter* **2006**, *20*, 119–124. [CrossRef] [PubMed]
39. Kumachev, A.; Tumarkin, E.; Walker, G.C.; Kumacheva, E. Characterization of the mechanical properties of microgels acting as cellular microenvironments. *Soft Matter* **2013**, *9*, 2959–2965. [CrossRef]
40. Mohapatra, H.; Kruger, T.M.; Lansakara, T.I.; Tivanski, A.V.; Stevens, L.L. Core and surface microgel mechanics are differentially sensitive to alternative crosslinking concentrations. *Soft Matter* **2017**, *13*, 5684–5695. [CrossRef] [PubMed]
41. Burmistrova, A.; Richter, M.; Uzum, C.; von Klitzing, R. Effect of cross-linker density of P(NIPAM-*co*-AAc) microgels at solid surfaces on the swelling/shrinking behaviour and the Young's modulus. *Colloid Polym. Sci.* **2011**, *289*, 613–624. [CrossRef]
42. Guo, M.Y.; Wyss, H.M. Micromechanics of Soft Particles. *Macromol. Mater. Eng.* **2011**, *296*, 223–229. [CrossRef]
43. Evans, E.A. Structure and Deformation Properties of Red Blood-Cells—Concepts and Quantitative Methods. *Method Enzymol.* **1989**, *173*, 3–35.
44. Guilak, F.; Tedrow, J.R.; Burgkart, R. Viscoelastic properties of the cell nucleus. *Biochem. Biophys. Res. Commun.* **2000**, *269*, 781–786. [CrossRef] [PubMed]
45. Hochmuth, R.M. Micropipette aspiration of living cells. *J. Biomech.* **2000**, *33*, 15–22. [CrossRef]
46. Sakai, A.; Murayama, Y.; Fujiwara, K.; Fujisawa, T.; Sasaki, S.; Kidoaki, S.; Yanagisawa, M. Increasing elasticity through changes in the secondary structure of gelatin by gelation in a microsized lipid space. *ACS Cent. Sci.* **2018**, in press. [CrossRef]
47. Wyss, H.M.; Franke, T.; Mele, E.; Weitz, D.A. Capillary micromechanics: Measuring the elasticity of microscopic soft objects. *Soft Matter* **2010**, *6*, 4550–4555. [CrossRef]
48. Kong, T.T.; Wang, L.Q.; Wyss, H.M.; Shum, H.C. Capillary micromechanics for core-shell particles. *Soft Matter* **2014**, *10*, 3271–3276. [CrossRef] [PubMed]
49. Kaufman, G.; Boltyanskiy, R.; Nejati, S.; Thiam, A.R.; Loewenberg, M.; Dufresne, E.R.; Osuji, C.O. Single-step microfluidic fabrication of soft monodisperse polyelectrolyte microcapsules by interfacial complexation. *Lab Chip* **2014**, *14*, 3494–3497. [CrossRef] [PubMed]
50. Mercadé-Prieto, R.; Nguyen, B.; Allen, R.; York, D.; Preece, J.A.; Goodwin, T.E.; Zhang, Z. Determination of the elastic properties of single microcapsules using micromanipulation and finite element modeling. *Chem. Eng. Sci.* **2011**, *66*, 2042–2049. [CrossRef]
51. Tagit, O.; Tomczak, N.; Vancso, G.J. Probing the morphology and nanoscale mechanics of single poly(*N*-isopropylacrylamide) microgels across the lower-critical-solution temperature by atomic force microscopy. *Small* **2008**, *4*, 119–126. [CrossRef] [PubMed]
52. Ao, M.; Xu, G.; Kang, W.; Meng, L.; Gong, H.; Zhou, T. Surface rheological behavior of gelatin/ionic liquid-type imidazolium gemini surfactant mixed systems. *Soft Matter* **2011**, *7*, 1199–1206. [CrossRef]

gels

MDPI

Article

Effect of Monomer Sequence along Network Chains on Thermoresponsive Properties of Polymer Gels

Shohei Ida *, Toru Kawahara, Hidekazu Kawabata, Tatsuya Ishikawa and Yoshitsugu Hirokawa *

Department of Materials Science, The University of Shiga Prefecture, 2500 Hassaka, Hikone, Shiga 522-8533, Japan; uspmatpolyst1@gmail.com (T.K.); uspmatpolyst2@gmail.com (H.K.); uspmatpolyst3@gmail.com (T.I.)

* Correspondence: ida.s@mat.usp.ac.jp (S.I.); hirokawa.yo@office.usp.ac.jp (Y.H.); Tel.: +81-749-28-8359 (S.I.); +81-749-28-8201 (Y.H.)

Received: 21 February 2018; Accepted: 9 March 2018; Published: 10 March 2018

Abstract: The effect of monomer sequence along the network chain on the swelling behavior of polymer gels should be clarified for the advanced control of swelling properties of gel materials. To this end, we systematically investigated the swelling properties of poly(acrylamide derivative) gels with the same composition but different monomer sequence by utilizing two gel synthetic methods: copolymerization giving a random network and co-crosslinking giving a blocky network. Both of the copolymerization and the co-crosslinking gels were prepared from the combination of two of the three following monomers: hydrophilic N,N-dimethylacrylamide (DMAAm), hydrophobic N-n-butylacrylamide (NBAAm), and thermoresponsive N-isopropylacrylamide (NIPAAm) with various monomer compositions. The swelling measurement of the obtained gels showed totally different behaviors between the copolymerization and the co-crosslinking gels, even with the same monomer composition. The copolymerization gels had the average property from the two monomers, depending on monomer composition, because random monomer distribution changed the affinity of each network chain to water. On the other hand, the co-crosslinking gels behaved as if two components independently contributed to the swelling properties, probably due to the domain structure derived from two kinds of prepolymers.

Keywords: gel; thermoresponsive property; monomer sequence; co-crosslinking; copolymerization; acrylamide derivative; swelling; volume phase transition

---

1. Introduction

Since the discovery of volume phase transition of polymer gels by Tanaka et al. [1–3], research on stimuli-responsive gels has been actively conducted. The most representative example is poly(N-isopropylacrylamide) (PNIPAAm) gel, which shows phase transition at around 33 °C in pure water and is anticipated to be useful in various applications [3–9]. In order to develop highly functional thermoresponsive materials, it is important to precisely control the thermoresponsive properties, including transition temperature, range of volume change, and transition speed, according to the purpose. The thermoresponsive property in water originates from the balance of hydrophilicity and hydrophobicity in the network chains: the former dominates the hydration at low temperature and the latter affects the aggregation of hydrophobic groups at high temperature. Therefore, an appropriate design of hydrophilicity/hydrophobicity balance in the structure is required for the control of thermoresponsive properties.

A general approach to control the swelling properties of polymer gels is the combination of plural monomers by copolymerization. Copolymerization of a monomer which gives a thermoresponsive gel such as N-isopropylacrylamide (NIPAAm) with another monomer easily creates a shift of the transition temperature due to changing the balance of hydrophilicity and hydrophobicity of network chains by feed ratio of the monomers [10–15]. Moreover, we have recently found that the adequate combination of hydrophilic monomer and hydrophobic monomer, both of which could not give a thermoresponsive polymer alone, could produce thermoresponsive gels showing sharp transition [16]. Thus, copolymerization of two kinds of monomers is a versatile method to control the thermoresponsive properties by regulating the balance of hydrophilicity and hydrophobicity.

Importantly, when combining two types of monomers, the gel properties should be strongly affected by the monomer sequence in the network chains, as the properties of linear polymers are greatly different between random sequence and block sequence. This indicates that a detailed understanding of the effect of monomer sequence is required for the advanced control of the swelling properties of the gels. However, commonly used radical copolymerization often produces only random sequences of two monomers in the network as shown in Figure 1a.

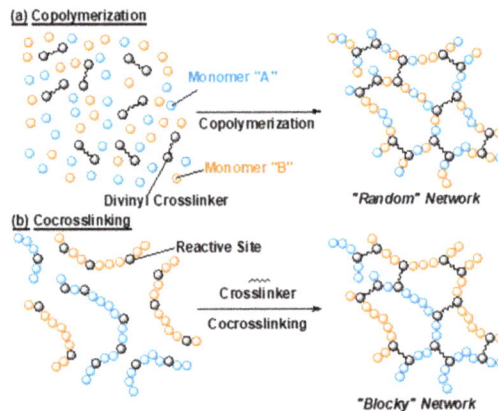

Figure 1. Schematic illustrations of gel network structure consisting of two kinds of monomer prepared by (**a**) copolymerization and (**b**) co-crosslinking.

On the other hand, it is known that a so-called amphiphilic conetwork (APCN) exhibits a characteristic swelling behavior. APCN has a structure combining two kinds of polymers, and shows the swelling behavior reflecting the properties of the two polymers [17,18]. For example, the combination of hydrophilic and hydrophobic polymers gives an APCN which can swell both in water and organic solvents [19–24]. Thermoresponsive polymers can be used as a constituent polymer of APCN, and the combination of thermoresponsive polymers and other polymers affords remarkable effects to the gels, such as improvement of response speed and thermoresponsive toughening [25–34]. Such an APCN structure is regarded as a contrasting sequence to a copolymerization gel having a random sequence. Therefore, systematic analysis of APCN and copolymerization gels using the same monomer combination would lead to understanding of the swelling behavior of the gel composed of two types of monomers, and it should contribute to the design criteria for the swelling properties of polymer gel materials.

In this paper, we focus on the effect of monomer sequence along network chains on the swelling properties of poly(acrylamide derivative) gels. To this end, the gels with the same monomer composition but a different monomer sequence along network chains were prepared by utilizing two synthetic methods. One method was copolymerization of two kinds of monomers to afford the *random* network in which each polymer chain between crosslinking points is a random copolymer like "-ABBABAAB-" (A and B stand for monomer units) as shown in Figure 1a. For the other method, to prepare the APCN structure, we utilized post-polymerization crosslinking methods using activated-ester chemistry [35]. Two kinds of poly(acrylamide derivative)s were employed for the crosslinking, and this "co-crosslinking" produces a *blocky* network in which all the network chains between crosslinking points are homopolymer like "-AAAAAA-" or "-BBBBBB-" (Figure 1b). We prepared copolymerization and co-crosslinking gels in various combination using three kinds of monomers with different affinity to water, which were hydrophilic, hydrophobic, and thermoresponsive, respectively. The swelling behavior in water of these gels with the same composition but different sequence was systematically investigated to elucidate the effect of network chain sequence on the thermoresponsive properties of the gels. It revealed that the copolymerization gels showed the averaged properties of the employed monomers and produced a change in the response temperature, while the co-crosslinking gels behaved as if the two components independently functioned and afforded a change in the swelling degree while keeping a constant response temperature.

2. Results and Discussion

2.1. Gel Synthesis by Copolymerization and Co-crosslinking

In order to clarify the effect of monomer sequence along network chains on swelling behavior of gels, copolymerization and co-crosslinking gels were prepared in various combinations using three kinds of monomers, as shown in Figure 2. These monomers give polymers with different affinity to water; hydrophilic *N*,*N*-dimethylacrylamide (DMAAm), hydrophobic *N*-*n*-butylacrylamide (NBAAm), and thermoresponsive NIPAAm.

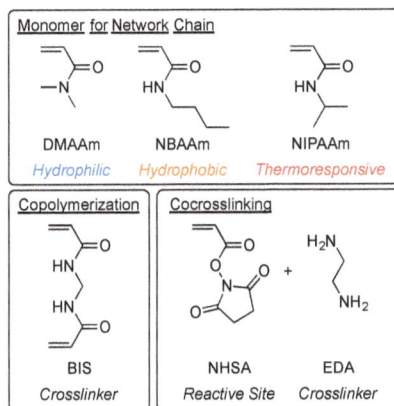

Figure 2. Chemical structure of monomers and crosslinkers for the synthesis of copolymerization and cocrosskinking gels. DMAAm = *N*,*N*-dimethylacrylamide; NBAAm = *N*-*n*-butylacrylamide (NBAAm); NIPAAm = *N*-isopropylacrylamide; BIS = *N*,*N*'-methylenebisacrylamide; NHSA = *N*-(acryloyloxy) succinimide; EDA = ethylenediamine.

Copolymerization gels were synthesized via radical copolymerization of two kinds of the acrylamide derivatives in the presence of the divinyl crosslinker, *N*,*N*'-methylenebisacrylamide (BIS), with various monomer feed ratios. The copolymerization gels were supposed to possess a random monomer sequence in the network chains because the structural similarity of the monomers gives the same polymerization reactivity. The reaction was conducted in methanol as solvent due to cosolvency, and the reaction time was set for 24 h, which was supposed to be long enough for the completion of the polymerization. Therefore, we could regard the feed ratio as the composition of the copolymerization gels. After the completion of gelation, methanol in the gel was replaced by pure water by means of immersion of the gels into large amounts of water for several days.

Co-crosslinking gels were obtained by a crosslinking reaction of two kinds of prepolymers containing activated ester groups, which react efficiently with primary amines [36]. The co-crosslinking gels are supposed to have a blocky network sequence because of the synthetic procedure using two kinds of polymers. The prepolymers were prepared by radical copolymerization of each acrylamide derivative monomer with the monomer carrying activated ester, *N*-(acryloyloxy)succinimide (NHSA). The obtained polymers were characterized by size-exclusion chromatography (SEC) and ^{1}H NMR analyses. The charactirization clarified that all the prepolymers had similar molecular weight and NHSA content as shown in Table 1. Then, the two kinds of prepolymers were dissolved in *N*,*N*-dimethylformamide (DMF), which is a good solvent for all the prepolymers, at various ratios to change the composition of the gels, and were then reacted with ethylenediamine (EDA) as crosslinker. After the completion of gelation, DMF in the gel was replaced by pure water in the same process used for the copolymerization gels. Herein, it should be noted that the copolymerization gels and the co-crosslinking gels were prepared at the same feed concentration of monomer unit and crosslinker, which gave almost the same crosslinking density under the assumption that all monomers and crosslinkers were completely consumed.

Table 1. Prepolymers for co-crosslinking gels prepared by radical copolymerization with NHSA. [1]

Prepolymer	NHSA Content (%) [2]	M_n [3]	M_w/M_n [3]
PDMAAm	7.7	43,900	1.78
PNBAAm	6.7	53,900	1.71
PNIPAAm	6.4	76,700	1.50

[1] The polymerization was conducted with the condition below: [acrylamide derivative] = 1900 mM, [NHSA] = 100 mM, [azobisisobutyronitrile (AIBN)] = 20 mM in 1,4-dioxane at 60 °C. [2] Determined by ^{1}H NMR analysis. [3] Determined by size-exclusion chromatography (SEC) analysis with poly(methyl methacrylate) standard. PDMAAm = poly(*N*-dimethylacrylamide); PNBAAm = poly(*N*-*n*-butylacrylamide); PNIPAAm = poly(*N*-isopropylacrylamide).

2.2. Hydrophilic/Hydrophobic Combination

The swelling properties of the copolymerization and the co-crosslinking gels with a variety of compositions were systematically compared. First, the combination of hydrophilic DMAAm and hydrophobic NBAAm was examined. The copolymerization gels and the co-crosslinking gels were different in their appearance as shown in Figure 3. The copolymerization gels were transparent in water at room temperature even with the NBAAm contents in the network chains reaching 50% (Figure 3b). On the contrary, the co-crosslinking gels showed turbidity during solvent replacement by immersion into water even with hydrophobic NBAAm composition as low as 10% (Figure 3c).

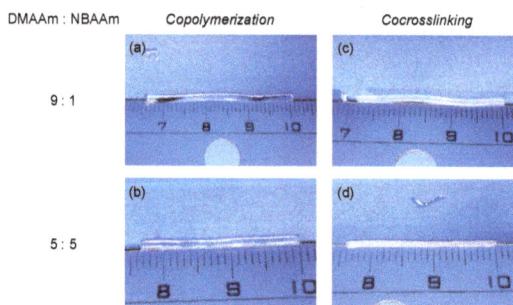

Figure 3. Appearance of (**a**,**b**) copolymerization and (**c**,**d**) co-crosslinking gels obtained from DMAAm and NBAAm with various monomer unit composition; DMAAm:NBAAm = (**a**,**c**) 9:1 and (**b**,**d**) 5:5.

This difference in appearance was supposed to be derived from the change of the network structure. Focusing on the hydrophobic moieties, NBAAm monomer units randomly distributed in the network of the copolymerization gels, while PNBAAm blocks were introduced in the co-crosslinking gels due to the difference in the preparation method. Therefore, the hydrophobicity of NBAAm units in the co-crosslinking gels affected the swelling state in water much more strongly than did that of the copolymerization gels with the same composition. That is, the aggregation of hydrophobic blocks easily occurred in the co-crosslinking gels. When this aggregation reached the size of the wavelength of the visible light, the turbidity was observed. In the co-crosslinking gels, this aggregation occurred even with low NBAAm content.

Then, the swelling behavior against temperature change of the obtained gels was investigated. Here, the degree of swelling was determined as the ratio of the volume at each temperature against the initial volume at the preparation state. We have recently reported that the copolymerization gel of hydrophilic and hydrophobic acrylamide derivatives with appropriate monomer ratio showed thermoresponsive properties with a sharp volume change in water [16]. The combination of DMAAm and NBAAm was the typical example. Increasing NBAAm content of the copolymerization gel gave a large decrease in the swelling degree against temperature rising, as shown in Figure 4a. Particularly, the gels with DMAAm:NBAAm = 5:5 showed sharp shrinking.

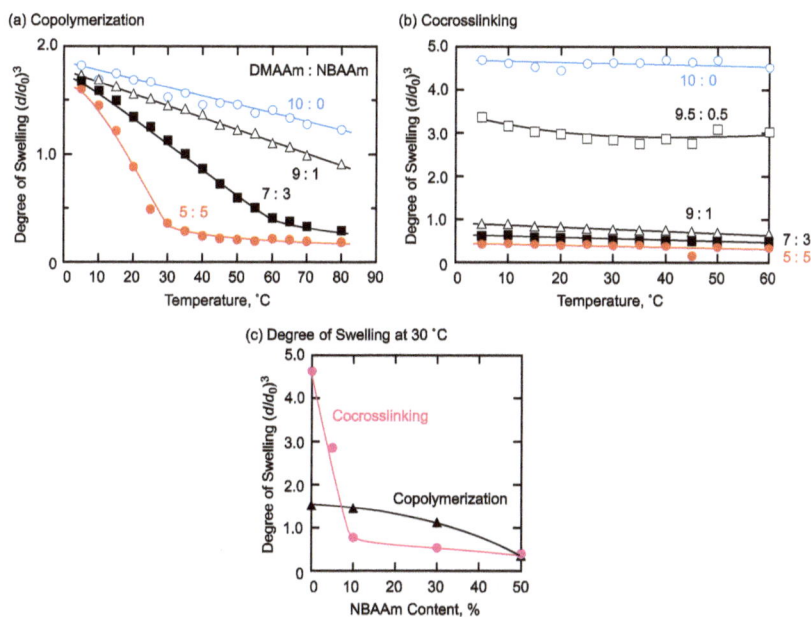

Figure 4. Swelling behavior of DMAAm/NBAAm (**a**) copolymerization and (**b**) co-crosslinking gels; (**c**) Relation between NBAAm content and the degree of swelling at 30 °C. Preparation condition of copolymerization gels: [DMAAm] + [NBAAm] = 1400 mM, [BIS] = 48.6 mM, [AIBN] = 55.6 mM in methanol at 55 °C. Preparation condition of co-crosslinking gels: [DMAAm unit] + [NBAAm unit] = 1400 mM, [NHSA unit] = [amino groups in EDA] in *N,N*-dimethylformamide (DMF) at room temperature.

In contrast, the co-crosslinking gels showed different behavior from that of the copolymerization gels as shown in Figure 4b. Introduction of NBAAm blocks to the co-crosslinking gels led to the decrease in the degree of swelling irrespective of the NBAAm content, and all composition of the gels did not give any change in the degree of swelling against temperature change. Here, there was a large difference between the degree of swelling of the copolymerization and that of the co-crosslinking gels with DMAAm:NBAAm = 10:0. This was probably due to the difference in the crosslinking efficiency in both methods. We employed the same monomer/crosslinker concentration at the two preparation systems, but the crosslinking efficiency was difficult to strictly control at this stage. Particularly, the solubility of the polymer to the reaction solvent was considered to affect the gelation behavior in the co-crosslinking reaction. As shown in the next section, such differences were not observed in the case of PNIPAAm gel. Thus, the solvent effect on the swelling degree appears in the case of PDMAAm co-crosslinking gel.

The difference in the swelling behavior of the copolymerization and the co-crosslinking gels was also clearly observed in the relation of NBAAm content and the degree of swelling at 30 °C as shown in Figure 4c. In the case of the copolymerization gels, the degree of swelling gradually decreased as the NBAAm content increased. On the other hand, the degree of swelling of the co-crosslinking gels sharply decreased at around 5~10% NBAAm content.

These results could be interpreted as the effect of the network chain sequence of the copolymerization and the co-crosslinking gels. In the copolymerization gels, hydrophilic and hydrophobic monomer units were randomly distributed along the network chains, and an increase in the content of NBAAm afforded alternating structures of DMAAm and NBAAm in the network chains [16]. This structure is similar to PNIPAAm gels in which the hydrophilic amide group and hydrophobic isopropyl group adjoined in one monomeric unit to realize an adequate balance of hydrophilicity and hydrophobicity for thermoresponsiveness. Similarly, the adequate balance of hydrophilicity/hydrophobicity was realized in hydrophilic/hydrophobic copolymerization gels to show thermoresponsive swelling behavior. On the other hand, since each monomer unit was separately introduced in the network in the case of the co-crosslinking gels, the swelling behavior was determined by the balance of hydrophilicity of PDMAAm and hydrophobicity of PNBAAm. Furthermore, this blocky sequence easily induced the aggregation of hydrophobic blocks in water to cause a decrease in the degree of swelling. The effect of hydrophobicity of PNBAAm became superior to the hydrophilic effect of PDMAAm at around 10% NBAAm content, and the sharp decrease of the degree of swelling was observed in this region. The blocky sequence also prevented the formation of amphiphilic local structures like copolymerization gels, and, therefore, the co-crosslinking gels did not show any thermoresponsiveness. Thus, the difference in network chain sequence strongly affected the swelling behavior of the gel prepared as a combination of hydrophilic and hydrophobic monomers.

2.3. Thermoresponsive/Hydrophobic Combination

The difference in the network chain sequence is supposed to strongly affect the thermoresponsive property of the gels, because the sequence relates to the hydration structure. Then, the effect of network chain sequence on the combination of thermoresponsive and hydrophobic monomers was examined by employing NIPAAm and NBAAm for copolymerization and co-crosslinking gel synthesis. Both synthetic methods produced gels showing shrinkage in response to temperature change in water, but increasing hydrophobic units had different effects on the swelling behavior, as shown in Figure 5.

As shown in Figure 5a, incorporation of hydrophobic NBAAm by copolymerization led to a lower shrinking temperature than that of PNIPAAm homopolymer gels until the NBAAm content reached 30%. This tendency was reasonable as it was previously observed in the case of thermoresponsive polymers obtained by copolymerization of NIPAAm with hydrophobic monomers [11]. The thermoresponsive property of PNIPAAm gel strongly relates to the balance of hydrophilicity and hydrophobicity. That is, hydration by amide groups leads to absorption and retention of water in the network at lower temperatures. Increasing temperature weakens this hydration effect, and aggregation by hydrophobic isopropyl groups induced volume collapse of the gels. In the NIPAAm/NBAAm copolymerization gel, randomly incorporated hydrophobic monomers suppressed the hydration of the network and strengthened hydrophobic interactions by butyl groups, resulting in a decreased shrinking temperature compared to that of the PNIPAAm gel. Notably, the copolymerization gel with NIPAAm:NBAAm = 5:5 seemed to show higher shrinking temperature than the gel with NIPAAm:NBAAm = 7:3. However, this gel was in the shrunken state even at 5 °C, and it showed turbidity and deformed at around 15 °C. After that, the macroscopic volume change of this gel may not follow the temperature change, and the apparent shrinking temperature would become higher.

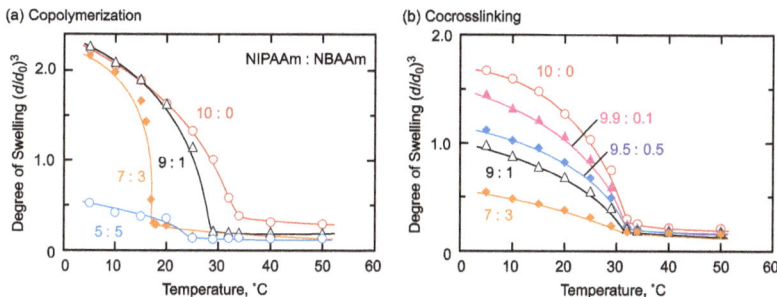

Figure 5. Swelling behavior of NIPAAm/NBAAm (**a**) copolymerization and (**b**) co-crosslinking gels. Preparation condition of copolymerization gels: [NIPAAm] + [NBAAm] = 1400 mM, [BIS] = 48.6 mM, [AIBN] = 55.6 mM in methanol at 55 °C. Preparation condition of co-crosslinking gels: [NIPAAm unit] + [NBAAm unit] = 1400 mM, [NHSA unit] = [amino groups in EDA] in DMF at room temperature.

On the other hand, the co-crosslinking gels showed the same transition temperature at around 32 °C regardless of the hydrophobic monomer composition, while the degree of swelling decreased with increasing NBAAm content as shown in Figure 5b. This suggests the presence of the domain structure in the network, which separately and independently comported. The co-crosslinking gels were prepared by a post-polymerization crosslinking reaction of two kinds of prepolymers. This preparation method produced a gel in which each network chain between the crosslinking points consisted of either PNIPAAm or PNBAAm homopolymer. This "blocky" sequence formed thermoresponsive and hydrophobic domains derived from each homopolymer chain. As indicated above, the thermoresponsive property of gels is derived from the affinity of network chains to water. Therefore, the same shrinking temperature was supposed to be due to the presence of PNIPAAm domains in NIPAAm/NBAAm co-crosslinking gels, because the affinity to water of thermoresponsive moieties in the co-crosslinking gels was equal to that of PNIPAAm homopolymer gels regardless of the presence of hydrophobic monomer units. At the same time, the hydrophobic domains in the co-crosslinking gels was attributed to the degree of swelling by aggregation in water.

Thus, the monomer sequence of network chains strongly affects the swelling properties of thermoresponsive gel's incorporated hydrophobic moieties. Randomly distributed NBAAm units contributed to the decrease of transition temperature, while hydrophobic block structure affected the degree of swelling without change of the transition temperature.

2.4. Thermoresponsive/Hydrophiliic Combination

A similar phenomenon to that of the thermoresponsive/hydrophobic combination was observed in the case of the thermoresponsive/hydrophilic combination. The copolymerization and the co-crosslinking gels from thermoresponsive NIPAAm and hydrophilic DMAAm were prepared, and the temperature dependence of the degree of swelling was measured in pure water (Figure 6). The copolymerization gels showed an increase in the shrinking temperature as the DMAAm content in the network increased (Figure 6a). On the other hand, the transition temperature of the co-crosslinking gels (NIPAAm:DMAAm = 7:3, 5:5, 3:7) was almost the same but higher than that of the PNIPAAm homopolymer gel (Figure 6b).

This difference of shrinking behavior could be also attributed to the difference in monomer sequences of the network chains, similar to that in NIPAAm/NBAAm gels. The randomly distributed DMAAm in the copolymerization gels strengthened hydrophilicity of the network, and led to the increase of transition temperature. On the other hand, DMAAm and NIPAAm units were incorporated in a blocky structure in the co-crosslinking gels, and the thermoresponsive domains could work independently. This led to almost the same transition temperature in the co-crosslinking gels. However, hydrophilic polymers affected the hydration structure of the gel, and the transition temperature was varied to be higher than that of the PNIPAAm homopolymer gel. Moreover, increasing hydrophilic moieties in the co-crosslinking gels simply produced an increase in the degree of swelling throughout the examined temperature.

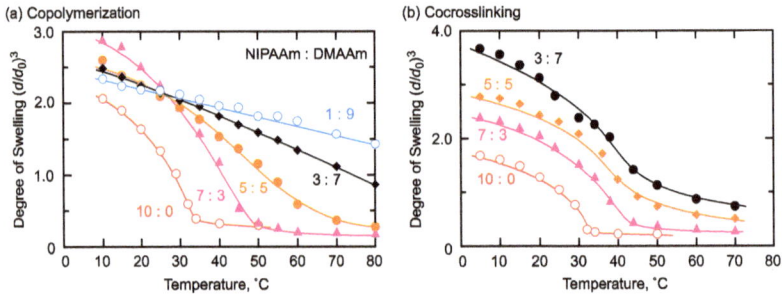

Figure 6. Swelling behavior of NIPAAm/DMAAm (**a**) copolymerization and (**b**) co-crosslinking gels. Preparation condition of copolymerization gels: [NIPAAm] + [DMAAm] = 1400 mM, [BIS] = 48.6 mM, [AIBN] = 55.6 mM in methanol at 55 °C. Preparation condition of co-crosslinking gels: [NIPAAm unit] + [DMAAm unit] = 1400 mM, [NHSA unit] = [amino groups in EDA] in DMF at room temperature.

Furthermore, the sequence effects on the shrinking rate was evaluated. The copolymerization and the co-crosslinking gels with the composition of NIPAAm:DMAAm = 7:3 were immersed into cold water at 5 °C to reach the equilibrium swelling state. Then, the gels were quickly transferred to 50 °C hot water, and the time-dependent change of the degree of swelling was observed. As shown in Figure 7a, both gels smoothly shrunk just after the elevation of temperature without an induction period, but the co-crosslinking gel shrunk much more rapidly than did the copolymerization gel.

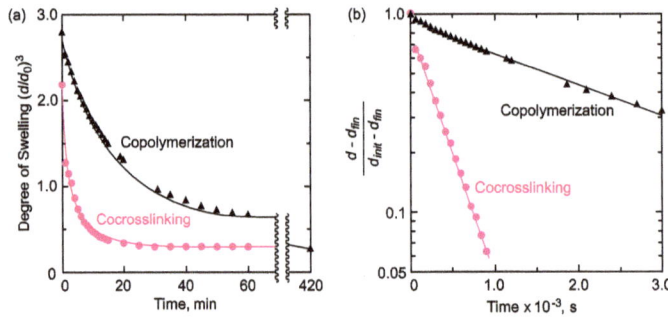

Figure 7. (**a**) Shrinking behavior of the copolymerization and the co-crosslinking gels of NIPAAm and DMAAm (7:3) against a temperature jump from 5 to 50 °C and (**b**) kinetic analysis of shrinking behavior upon temperature jump. When an isotropic shrinking is observed, the relaxation time (τ) can be determined from the slope of the line ($-1/\tau$)

Next, we performed a quantitative analysis of the shrinking kinetics of the co-crosslinking and the copolymerization gels. For the swelling/shrinking of cylinder-shaped gels, the time dependence of diameter change can be described as

$$\frac{d - d_{fin}}{d_{init} - d_{fin}} = \frac{6}{\pi^2} exp\left(-\frac{t}{\tau}\right) \tag{1}$$

where d_{init}, d_{fin}, and τ, respectively, stand for the equilibrium diameter at the initial swelling state, the equilibrium diameter at the shrunken state, and the relaxation time constant, when t is larger than τ and no phase separation occurs [37–39]. We applied this equation for the results of Figure 7a and determined τ for the co-crosslinking and the copolymerization gels: $\tau = 3.4 \times 10^2$ s for the co-crosslinking gel and $\tau = 2.5 \times 10^3$ s for the copolymerization gel (Figure 7b). For cylindrical gels, the relaxation time (τ) is related to the cooperative diffusion coefficient (D):

$$D = \frac{3d_{fin}^2}{8\pi^2\tau} \tag{2}$$

From this relationship, we estimated D at 8.6×10^{-7} cm^2·s^{-1} for the co-crosslinking gel and 1.1×10^{-7} cm^2·s^{-1} for the copolymerization gel. These values indicated that the co-crosslinking gel shrunk much faster than did the copolymerization gel. Furthermore, it can be considered that the dehydrated water spilled out of the gels much more smoothly in the case of the co-crosslinking gel, since the value D is related to the modulus of polymer chains and the frictional coefficient between the network and water.

This quick shrinking of the co-crosslinking gel is due to the presence of hydrophilic domains derived from PDMAAm chains. It has been reported that incorporation of hydrophilic polymers or domains into the thermoresponsive gel accelerates the shrinking rate because dehydrated water can effectively spill out of the gels through the hydrophilic domains [27–30,40–42]. The same mechanism could be applied to the PNIPAAm/PDMAAm co-crosslinking gels. Namely, PDMAAm domains functioned as water pathways to induce much more rapid shrinking of the co-crosslinking gel than that of the copolymerization gel, in which hydrophilic monomer units randomly distributed and domain structure was not formed.

3. Conclusions

We evaluated the effect of monomer sequence along the network chain of polymer gels on the swelling behavior by utilizing two gel synthetic methods: copolymerization and co-crosslinking. We clearly demonstrated that the copolymerization gels and the co-crosslinking gels showed different swelling behavior even with the same monomer composition in a series of monomer combinations including hydrophilic/hydrophobic, thermoresponsive/hydrophobic, and thermoresponsive/hydrophilic acrylamide derivatives. Importantly, the copolymerization method afforded random monomer distribution to change the affinity of each network chain to water depending on monomer composition. On the other hand, the co-crosslinking method produced a blocky sequence derived from two kinds of prepolymers, which separately and independently contributed to the swelling properties such as the degree of swelling and transition temperature. Thus, we clarified the effect of monomer sequence along the network chain on the properties of polymer gels. These fundamental results are important for the design of polymer gel materials such as stimuli-responsive materials and biomedical devices in combination with plural monomers.

4. Materials and Methods

4.1. Materials

DMAAm was kindly provided by Kohjin Co., Ltd. (Tokyo, Japan), and it was purified by distillation before use. NBAAm was prepared by the reaction of acryloyl chloride with *n*-butylamine, and purified by distillation before use. NIPAAm (Wako, Osaka, Japan, 98%) was purified by recrystallization from toluene/*n*-hexane. NHSA was prepared as reported in literature [43]. BIS (Wako, 99%), EDA (Wako, 98%), azobisisobutyronitrile (AIBN; Wako, 98%), 1,2,3,4-tetrahydronaphthalene (tetralin; Aldrich, St. Louis, MI, USA, 99%), 1,4-dioxane (Wako, 99%), DMF (Wako, 99%), and methanol (Wako) were used as received.

4.2. Copolymerization Gel Synthesis

A typical example of gel synthesis by copolymerization is given below. DMAAm (0.243 g), NBAAm (0.312 g), and BIS (52.5 mg) were all dissolved into methanol (7.0 mL). After addition of AIBN (63.9 mg), the solution was transferred into a test tube containing glass capillaries (volume: 40 μL, internal diameter: 1.3 mm) and bubbled with nitrogen for 10 min. The test tube was immersed in the water bath controlled at 55 °C and kept for 24 h for the completion of the gelation. Under these conditions, almost all of the monomers were consumed to become the gel, and the network polymer composition was considered to be equal to the monomer concentration ratio. Afterwards, the cylindrical gels were taken out from the capillaries and washed with distilled water by means of immersion for several days.

4.3. Co-crosslinking Gel Synthesis

A typical example of gel synthesis by co-crosslinking is given below. DMAAm (1.95 mL), NHSA (169 mg), tetralin (0.50 mL), AIBN (32.8 mg), and 1,4-dioxane (7.5 mL) were added to a 50 mL round-bottomed flask equipped with a three-way stopcock and bubbled with nitrogen for 10 min. The flask was then placed in an oil bath kept at 60 °C for 70 min. The reaction was quenched by cooling down the reaction solution to −78 °C. Monomer conversion was determined from the concentration of residual monomer measured by ^{1}H NMR with tetralin as an internal standard. Then, the reaction mixture was poured into diethyl ether to obtain the purified P(DMAAm/NHSA) prepolymers. The monomer composition of the polymers was determined by ^{1}H NMR analysis. PNBAAm and PNIPAAm prepolymers were also prepared by the same procedures.

The obtained PDMAAm prepolymer [NHSA content: 7.7 mol% (calculated by ^{1}H NMR)] (79.3 mg) and the PNBAAm prepolymer [NHSA content: 6.7 mol%] (95.4 mg) were dissolved into DMF (0.50 mL). To this polymer solution, 0.5 mL of EDA/DMF solution (amino groups were set to be equimolar amounts of NHSA units in the prepolymers) was added, and glass capillaries (volume: 40 μL, internal diameter: 1.3 mm) were added to the reaction vessel. The mixture was kept for 24 h to complete the crosslinking reaction. Then, the cylindrical gels were taken out from the capillaries and washed with distilled water by means of immersion for several days to remove the resultant *N*-hydroxysuccinimide, which was produced by the reaction with activated ester and diamine crosslinker.

4.4. Characterization

Number-average molecular weight (M_n) and polydispersity index (M_w/M_n) of polymers were determined by size-exclusion chromatography (SEC) in DMF containing 10 mM LiBr at 40 °C using three polystyrene gel columns (PLgel 5 μm MIXED-C, PLgel 3 μm MIXED-E and Shodex KF-805L) that were connected to a Shimadzu LC-10AD precision pump (Kyoto, Japan) and a Shimadzu RID-10A refractive index detector (Kyoto, Japan). The columns were calibrated against standard poly(methyl methacrylate) samples (Agilent, Santa Clara, California, USA). ^{1}H NMR spectra were recorded on a JEOL JNM-LA400 spectrometer (Akishima, Japan), operating at 399.65 MHz. The degree of swelling was determined by measurement of the diameter of the cylindrical gels. The gels were immersed

Gels **2018**, *4*, 22

in water at predetermined temperatures, and the equilibrium diameter at given temperature, d, was measured using a digital microscope (MOTICAM2000, Shimadzu, Kyoto, Japan). The degree of swelling was calculated by $(d/d_0)^3$; d_0 is the inner diameter of the capillary used for the gel preparation. The shrinking rate was evaluated by temperature-jump experiment. The cylindrical gels were immersed in water at 5 °C until the gels reached to the equilibrium swollen state. Then, the gels were quickly transferred into the hot water at 50 °C, and the change of the diameter of the gel was observed.

Acknowledgments: This work was partially supported by the Japan Society for the Promotion of the Science through a Grant-in-aid for Scientific Research (C) No. 24550259, for which the authors are grateful.

Author Contributions: Toru Kawahara, Hidekazu Kawabata, and Tatsuya Ishikawa performed the experiments and analyzed the data. Shohei Ida and Yoshitsugu Hirokawa designed the concepts and experiments, analyzed the data, and wrote manuscript.

Conflicts of Interest: The authors declare no conflict of interest.

References

1. Tanaka, T. Collapse of Gels and the Critical Endpoint. *Phys. Rev. Lett.* **1978**, *40*, 820–823. [CrossRef]
2. Tanaka, T.; Fillmore, D.; Sun, S.-T.; Nishio, I.; Swislow, G.; Shah, A. Phase Transitions in Ionic Gels. *Phys. Rev. Lett.* **1980**, *45*, 1636–1639. [CrossRef]
3. Hirokawa, Y.; Tanaka, T. Volume phase transition in a nonionic gel. *J. Chem. Phys.* **1984**, *81*, 6379–6380. [CrossRef]
4. Schild, H.G. Poly(*N*-isopropylacrylamide): Experiment, theory and application. *Prog. Polym. Sci.* **1992**, *17*, 163–249. [CrossRef]
5. Halperin, A.; Kröger, M.; Winnik, F.M. Poly(*N*-isopropylacrylamide) Phase Diagrams: Fifty Years of Research. *Angew. Chem. Int. Ed.* **2015**, *54*, 15342–15367. [CrossRef] [PubMed]
6. Osada, Y.; Gong, J. Stimuli-responsive polymer gels and their application to chemomechanical systems. *Prog. Polym. Sci.* **1993**, *18*, 187–226. [CrossRef]
7. Kikuchi, A.; Okano, T. Pulsatile drug release control using hydrogels. *Adv. Drug Deliv. Rev.* **2002**, *54*, 53–77. [CrossRef]
8. Chaterji, S.; Kwon, I.K.; Park, K. Smart polymeric gels: Redefining the limits of biomedical devices. *Prog. Polym. Sci.* **2007**, *32*, 1083–1122. [CrossRef] [PubMed]
9. Diaz Diaz, D.; Kuhbeck, D.; Koopmans, R.J. Stimuli-Responsive Gels as Reaction Vessels and Reusable Catalysts. *Chem. Soc. Rev.* **2011**, *40*, 427–448. [CrossRef] [PubMed]
10. Feil, H.; Bae, Y.H.; Feijen, J.; Kim, S.W. Effect of comonomer hydrophilicity and ionization on the lower critical solution temperature of N-isopropylacrylamide copolymers. *Macromolecules* **1993**, *26*, 2496–2500. [CrossRef]
11. Takei, Y.G.; Aoki, T.; Sanui, K.; Ogata, N.; Okano, T.; Sakurai, Y. Temperature-responsive bioconjugates. 2. Molecular design for temperature-modulated bioseparations. *Bioconjug. Chem.* **1993**, *4*, 341–346. [CrossRef] [PubMed]
12. Katakai, R.; Saito, K.; Sorimachi, M.; Hiroki, A.; Kinuno, T.; Nakajima, T.; Shimizu, M.; Kubota, H.; Yoshida, M. Hydrophobic Site-Specific Control of the Volume Phase Transition of Hydrogels. *Macromolecules* **1998**, *31*, 3383–3384. [CrossRef]
13. Lutz, J.-F.; Hoth, A. Preparation of Ideal PEG Analogues with a Tunable Thermosensitivity by Controlled Radical Copolymerization of 2-(2-Methoxyethoxy)ethyl Methacrylate and Oligo(ethylene glycol) Methacrylate. *Macromolecules* **2006**, *39*, 893–896. [CrossRef]
14. Park, J.-S.; Kataoka, K. Precise Control of Lower Critical Solution Temperature of Thermosensitive Poly(2-isopropyl-2-oxazoline) via Gradient Copolymerization with 2-Ethyl-2-oxazoline as a Hydrophilic Comonomer. *Macromolecules* **2006**, *39*, 6622–6630. [CrossRef]
15. Confortini, O.; Du Prez, F.E. Functionalized Thermo-Responsive Poly(vinyl ether) by Living Cationic Random Copolymerization of Methyl Vinyl Ether and 2-Chloroethyl Vinyl Ether. *Macromol. Chem. Phys.* **2007**, *208*, 1871–1882. [CrossRef]

16. Ida, S.; Kawahara, T.; Fujita, Y.; Tanimoto, S.; Hirokawa, Y. Thermoresponsive Properties of Copolymer Gels Induced by Appropriate Hydrophilic/Hydrophobic Balance of Monomer Combination. *Macromol. Symp.* **2015**, *350*, 14–21. [CrossRef]

17. Patrickios, C.S.; Georgiou, T.K. Covalent amphiphilic polymer networks. *Curr. Opin. Colloid Interface Sci.* **2003**, *8*, 76–85. [CrossRef]

18. Erdodi, G.; Kennedy, J.P. Amphiphilic conetworks: Definition, synthesis, applications. *Prog. Polym. Sci.* **2006**, *31*, 1–18. [CrossRef]

19. Weber, M.; Stadler, R. Hydrophilic-hydrophobic two-component polymer networks: 1. Synthesis of reactive poly(ethylene oxide) telechelics. *Polymer* **1988**, *29*, 1064–1070. [CrossRef]

20. Weber, M.; Stadler, R. Hydrophilic-hydrophobic two-component polymer networks: 2. Synthesis and characterization of poly(ethylene oxide)-linked-polybutadiene. *Polymer* **1988**, *29*, 1071–1078. [CrossRef]

21. Simmons, M.R.; Yamasaki, E.N.; Patrickios, C.S. Cationic Amphiphilic Model Networks: Synthesis by Group Transfer Polymerization and Characterization of the Degree of Swelling. *Macromolecules* **2000**, *33*, 3176–3179. [CrossRef]

22. Triftaridou, A.I.; Hadjiyannakou, S.C.; Vamvakaki, M.; Patrickios, C.S. Synthesis, Characterization, and Modeling of Cationic Amphiphilic Model Hydrogels: Effects of Polymer Composition and Architecture. *Macromolecules* **2002**, *35*, 2506–2513. [CrossRef]

23. Erdodi, G.; Kennedy, J.P. Ideal tetrafunctional amphiphilic PEG/PDMS conetworks by a dual-purpose extender/crosslinker. I. Synthesis. *J. Polym. Sci. Part A Polym. Chem.* **2005**, *43*, 4953–4964. [CrossRef]

24. Erdodi, G.; Kennedy, J.P. Ideal tetrafunctional amphiphilic PEG/PDMS conetworks by a dual-purpose extender/crosslinker. II. Characterization and properties of water-swollen membranes. *J. Polym. Sci. Part A Polym. Chem.* **2005**, *43*, 4965–4971. [CrossRef]

25. Guan, Y.; Ding, X.; Zhang, W.; Wan, G.; Peng, Y. Polytetrahydrofuran Amphiphilic Networks, 5. Synthesis and Swelling Behavior of Thermosensitive Poly(N-isopropylacrylamide)-L-polytetrahydrofuran Networks. *Macromol. Chem. Phys.* **2002**, *203*, 900–908. [CrossRef]

26. Lequieu, W.; Du Prez, F.E. Segmented polymer networks based on poly(N-isopropyl acrylamide) and poly(tetrahydrofuran) as polymer membranes with thermo-responsive permeability. *Polymer* **2004**, *45*, 749–757. [CrossRef]

27. Hirotsu, S. Anomalous Kinetics of the Volume Phase Transition in Poly-N-Isopropylacrylamide Gels. *Jpn. J. Appl. Phys.* **1998**, *37*, L284. [CrossRef]

28. Kaneko, Y.; Nakamura, S.; Sakai, K.; Aoyagi, T.; Kikuchi, A.; Sakurai, Y.; Okano, T. Rapid Deswelling Response of Poly(N-isopropylacrylamide) Hydrogels by the Formation of Water Release Channels Using Poly(ethylene oxide) Graft Chains. *Macromolecules* **1998**, *31*, 6099–6105. [CrossRef]

29. Zheng, Q.; Zheng, S. From poly(N-isopropylacrylamide)-block-poly(ethylene oxide)-block-poly(N-isopropylacrylamide) triblock copolymer to poly(N-isopropylacrylamide)-block-poly(ethylene oxide) hydrogels: Synthesis and rapid deswelling and reswelling behavior of hydrogels. *J. Polym. Sci. Part A Polym. Chem.* **2012**, *50*, 1717–1727. [CrossRef]

30. Kamata, H.; Chung, U.; Shibayama, M.; Sakai, T. Anomalous volume phase transition in a polymer gel with alternative hydrophilic-amphiphilic sequence. *Soft Matter* **2012**, *8*, 6876–6879. [CrossRef]

31. Kamata, H.; Akagi, Y.; Kayasuga-Kariya, Y.; Chung, U.-I.; Sakai, T. "Nonswellable" Hydrogel Without Mechanical Hysteresis. *Science* **2014**, *343*, 873–875. [CrossRef] [PubMed]

32. Guo, H.; Sanson, N.; Hourdet, D.; Marcellan, A. Thermoresponsive Toughening with Crack Bifurcation in Phase-Separated Hydrogels under Isochoric Conditions. *Adv. Mater.* **2016**, *28*, 5857–5864. [CrossRef] [PubMed]

33. Guo, H.; Mussault, C.; Brûlet, A.; Marcellan, A.; Hourdet, D.; Sanson, N. Thermoresponsive Toughening in LCST-Type Hydrogels with Opposite Topology: From Structure to Fracture Properties. *Macromolecules* **2016**, *49*, 4295–4306. [CrossRef]

34. Ida, S.; Kitanaka, H.; Ishikawa, T.; Kanaoka, S.; Hirokawa, Y. Swelling properties of thermoresponsive/hydrophilic co-networks with functional crosslinked domain structures. *Polym. Chem.* **2018**, in press. [CrossRef]

35. Ida, S.; Katsurada, A.; Yoshida, R.; Hirokawa, Y. Effect of reaction conditions on poly(N-isopropylacrylamide) gels synthesized by post-polymerization crosslinking system. *React. Funct. Polym.* **2017**, *115*, 73–80. [CrossRef]

36. Theato, P. Synthesis of well-defined polymeric activated esters. *J. Polym. Sci. Part A Polym. Chem.* **2008**, *46*, 6677–6687. [CrossRef]

37. Tanaka, T.; Fillmore, D.J. Kinetics of swelling of gels. *J. Chem. Phys.* **1979**, *70*, 1214–1218. [CrossRef]

38. Li, Y.; Tanaka, T. Kinetics of swelling and shrinking of gels. *J. Chem. Phys.* **1990**, *92*, 1365–1371. [CrossRef]

39. Hirose, H.; Shibayama, M. Kinetics of volume phase transition in poly(*N*-isopropylacrylamide-*co*-acrylic acid) gels. *Macromolecules* **1998**, *31*, 5336–5342. [CrossRef]

40. Zhang, X.-Z.; Xu, X.-D.; Cheng, S.-X.; Zhuo, R.-X. Strategies to improve the response rate of thermosensitive PNIPAAm hydrogels. *Soft Matter* **2008**, *4*, 385–391. [CrossRef]

41. Yan, H.; Fujiwara, H.; Sasaki, K.; Tsujii, K. Rapid Swelling/Collapsing Behavior of Thermoresponsive Poly(*N*-isopropylacrylamide) Gel Containing Poly(2-(methacryloyloxy)decyl phosphate) Surfactant. *Angew. Chem. Int. Ed.* **2005**, *44*, 1951–1954. [CrossRef] [PubMed]

42. Chen, X.; Tsujii, K. A Novel Hydrogel Showing Super-Rapid Shrinking but Slow Swelling Behavior. *Macromolecules* **2006**, *39*, 8550–8552. [CrossRef]

43. Pollak, A.; Blumenfeld, H.; Wax, M.; Baughn, R.L.; Whitesides, G.M. Enzyme immobilization by condensation copolymerization into crosslinked polyacrylamide gels. *J. Am. Chem. Soc.* **1980**, *102*, 6324–6336. [CrossRef]

gels

MDPI

Article

Process Variable Optimization in the Manufacture of Resorcinol–Formaldehyde Gel Materials

Martin Prostredný, Mohammed G. M. Abduljalil, Paul A. Mulheran and Ashleigh J. Fletcher *

Department of Chemical and Process Engineering, University of Strathclyde, Glasgow G1 1XJ, UK;
martin.prostredny@strath.ac.uk (M.P.); mohammed.abduljalil@strath.ac.uk (M.G.M.A.);
paul.mulheran@strath.ac.uk (P.A.M.)
* Correspondence: ashleigh.fletcher@strath.ac.uk; Tel.: +44-(0)-141-548-2431

Received: 15 December 2017; Accepted: 12 April 2018; Published: 17 April 2018

Abstract: Influence of process parameters of resorcinol–formaldehyde xerogel manufacture on final gel structure was studied, including solids content, preparation/drying temperature, solvent exchange, and drying method. Xerogels produced using a range of solids content between 10 and 40 w/v% show improved textural character up to 30 w/v% with a subsequent decrease thereafter. Preparation/drying temperature shows a minimal threshold temperature of 55 °C is required to obtain a viable gel structure, with minimal impact on gel properties for further thermal increase. Improving the solvent exchange method by splitting the same amount of acetone used in this phase over the period of solvent exchange, rather than in a single application, shows an increase in total pore volume and average pore diameter, suggesting less shrinkage occurs during drying when using the improved method. Finally, comparing samples dried under vacuum and at ambient pressure, there seems to be less shrinkage when using vacuum drying compared to ambient drying, but these changes are insubstantial. Therefore, of the process parameters investigated, improved solvent exchange seems the most significant, and it is recommended that, economically, gels are produced using a solids content of 20 w/v% at a minimum temperature of 55 °C, with regular solvent replenishment in the exchange step, followed by ambient drying.

Keywords: xerogel; Brunauer-Emmett-Teller theory; Barrett-Joyner-Halenda analysis; temperature; solids content; drying; solvent exchange

1. Introduction

Resorcinol–formaldehyde (RF) aerogels are a family of porous materials, first discovered in 1989 [1] by Pekala, and which have seen extensive application, due to their tailorable textural properties, in a range of applications, including as catalyst supports [2–4], in gas storage systems [5,6] and gas separation devices [7,8], in the fabrication of fuel cells [9,10], and as a core component in insulation [11,12]. The control of the porous character of these materials underpins their vast applicability, allowing tailored synthesis in terms of surface area, pore volume and pore size; however, the mechanism by which these gel materials form is not completely understood and there is significant scope for materials and process optimization.

It is generally accepted that the sol-gel polycondensation reaction of resorcinol (R) and formaldehyde (F) proceeds as shown in Figure 1; the reaction is also usually performed at above ambient temperatures. The reaction can be seen to proceed via an initial addition reaction between R and F, forming a hydroxymethyl derivative species, which undergoes self-condensation to create oligomeric chains that form clusters, and finally, a cross-linked 3D gel network. Our previous work, utilizing light scattering techniques, has provided insight into the mechanism of cluster growth, whereby, in a system with fixed reaction parameters, thermodynamics controls the size of growing clusters, while there is kinetic control of cluster population [13]. The reaction is promoted by the presence of a metal salt, known within the field as a catalyst. The most commonly used catalyst is sodium carbonate (Na_2CO_3), as originally used by Pekala, and the role of the metal carbonate is thought to be two-fold. While the carbonate is known to act as a base, promoting the initial reaction between resorcinol and formaldehyde through proton abstraction, the metal ion is thought to stabilise the colloidal suspension involved in development of clusters dispersed within the solvent matrix [14]. Hence, many studies have previously focused on the use of different catalytic species to control the final gel material [13–19]. However, it should be noted that the polycondensation reaction can also be influenced by a variety of other synthesis parameters, resulting in a modification of the porous character of the final aerogel product [20,21]. This includes synthesis parameters such as resorcinol to carbonate molar ratio (R/C) and the mass of solids dissolved within a fixed volume of solvent (deionised water) [22], as well as process variables, which can also affect the end material. Recent research has shown that both the time allowed for the reaction mixture to be stirred before heating [23], and the shape of the mould used to form the RF aerogel [24], can also have a significant effect on the internal structure of the gel product. The post-synthetic processing of RF gels is also subject to significant variation, in terms of solvent exchange and drying methods used, the former is usually selected to enhance the latter. Drying methods include supercritical drying, freeze drying or ambient temperature drying, with or without vacuum.

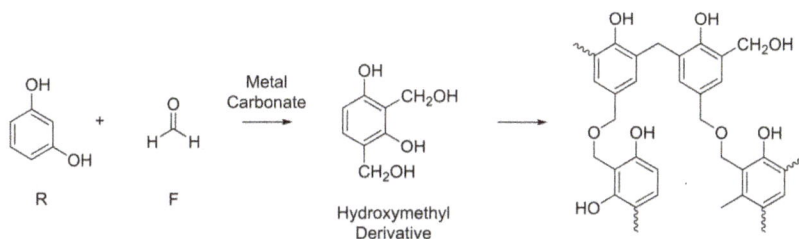

Figure 1. General reaction mechanism proposed in the reaction of resorcinol and formaldehyde. R: resorcinol; F: formaldehyde.

Kistler was instrumental in developing the first aerogels from silica based materials, and in his work, he had observed that evaporative drying results in destructive forces acting on the pore walls as a consequence of surface tension, and subsequent collapse of the gel [25]; he also established that, due to the high critical temperature and pressure of H_2O [26], and its poor solubility in supercritical solvents [27], the water entrained within the gel first had to be exchanged with a solvent that was completely miscible with the supercritical solvent. Following this work, the Lawrence Berkeley Laboratory [28] discovered that supercritical CO_2 could be used as a direct solvent replacement in the drying of silica aerogels [28], presenting a safer route to gel production. Pekala subsequently used this discovery, in conjunction with previous knowledge on RF resins to prepare organic aerogels [1,29]. Further studies, since then, discovered that other drying methods can be used, e.g., conventional evaporative drying to form xerogels [30] and freeze drying to form cryogels [31–34]. Czakkel et al. [32] compared the effects of evaporative drying in an inert atmosphere, freeze drying and supercritical drying, on the textural properties of RF gels, and found that the cryogels exhibited the highest pore volumes and surface areas due to the improved solvent quality of t-butanol; the evaporative samples showed less developed structures due to increased shrinkage arising from the formation of a liquid–vapour interface and resultant surface tension [20]. This indicates that the final drying step is critical to retention of porous character; however, Jabeen also demonstrated that exchanging entrained water with a solvent of lower surface tension reduced gel shrinkage and, as a result, increased pore volume [35]. The results indicate that, even in the event of a prolonged solvent exchange, residual water is retained within the pores of the gel, resulting in increased surface tension during drying, and impacting on the porous structure obtained. Another way to avoid liquid–vapour interfaces is to use freeze-drying [12–15]. It has been noted, in previous studies, that supercritical drying and freeze-drying are expensive to perform, and require specialist knowledge [34,36]; hence, a route to gel production that avoids such methods would be economically beneficial.

These previous works have established a base from which most researchers work to produce RF gels but, to date, there has been no overarching study that has investigated process optimization holistically, which is essential for the scaled production of these materials. Hence, the aim of this current work is to optimise synthesis parameters and process variables to provide tailored production of selected textural characteristics in the final material. This involves determination of the impact of the solvent exchange method, total solids content, and drying method used, with respect to with varying R/C ratio. This optimal system was then studied further by altering the temperature at which the steps of synthesis, curing and drying were all set, to determine the validity of the widely accepted temperature of 85 °C in the synthetic procedure, as this has potential impact on the basis of both economics and safety. Low temperature nitrogen sorption measurements were used to characterise the textural properties of the synthesised aerogels, allowing changes in the internal structure of the xerogel to be monitored and quantified.

2. Results and Discussion

2.1. Effect of Solvent Exchange Method

Gels, produced as outlined above, generally undergo solvent exchange for a period of three days with only an initial volume of acetone added to the drained, cured gel; however, this may not be the most appropriate method to retain the porous structure developed during synthesis. Due to the high surface tension value for water, over the synthetic temperature range used to produce RF gels, the process of drying hydrogels leads to significant shrinkage of the material, as a consequence of the resulting high stresses that act on the pore walls. Therefore, it is desirable to replace the water, entrained within the pores, with a liquid that exhibits a lower surface tension, and preferably a lower boiling point, than water, within the temperature range of interest. The surface tension of water is high, even at elevated temperatures, e.g., 67.94 N/m at 50 °C [37], and a number of alternative solvents, with reduced surface tensions e.g., amyl acetate, acetone, t-butanol and isopropanol [20,38], have been proposed for solvent exchange in previous studies; however, acetone offers an excellent combination of a reduction in surface tension (19.65 N/m at 50 °C [39]) and relatively low cost compared to alternative solvents. Hence, acetone was used for solvent exchange within this study.

Replacement of the liquid within the pores is driven by diffusion, although agitation is often used to enhance contact of the material and fresh solvent; hence, sufficient time is required for full exchange to occur. Another factor that is potentially important, in maximizing the level of exchange, is the water concentration gradient between the pore liquid and the bulk solvent surrounding the sample. To investigate the effect of the solvent exchange method used, three batches of gels, individually of 60 mL total liquid volume, were synthesised, each of which, after curing, were washed with acetone, drained and, subsequently, agitated in acetone for three days. The key difference was that the first two batches were used to investigate the effect of a different volume of acetone used in one application and were processed by adding the entire volume of acetone at the beginning of the three days, namely 180 or 240 mL, and the sample was left without further handling for the whole solvent exchange period, while the third batch was treated with a fresh volume of acetone each day for three successive days with the total volume of acetone used adding up to 240 mL, thus maintaining the same total volume of acetone as the second batch but splitting the total volume over multiple days.

The data obtained for the pore size distributions of the three batches of gels are shown in Figure 2, and it can be seen that changing the acetone bath daily has a more pronounced positive effect on the total pore volume of the RF gel samples compared to just increasing the total acetone volume without changing the bath daily, especially for samples with lower R/C ratios. Improving the solvent exchange method, by increasing the concentration gradient daily, leads to pores with larger average diameter

(Table 1). This, coupled with the increase in pore volume, is ascribed to a reduction in shrinkage during the drying stage. If the acetone bath is replaced daily, the water concentration gradient is renewed every day, thus there is an increased driving force, which removes more water from the pores. This leads to lower stresses being exerted on the pore walls during the drying stage, due to the lower surface tension of acetone compared to water. However, for samples with higher R/C ratios exhibiting a weaker gel structure, the improved method does not seem to have the same pronounced positive effect observed for the lower R/C gels with smaller average pore diameter. A possible explanation is that when the acetone bath is exchanged daily, the replenishment step slightly damages the softer structure, resulting in lower values of surface area and pore size. The findings from this section of work suggest that there is significant advantage in using an improved solvent exchange method for most of the samples, hence, all samples in the following sections were prepared using daily replenishment of acetone within the solvent exchange stage, with the intention of maintaining the gel structure as close to the original hydrogel structure as possible, without the need to use cryogenic or supercritical processing steps. It is important to note that, in order to obtain improved gel characteristics, it is not necessary to increase the amount of acetone used during the solvent exchange, rather it is imperative to split this amount over the exchange period.

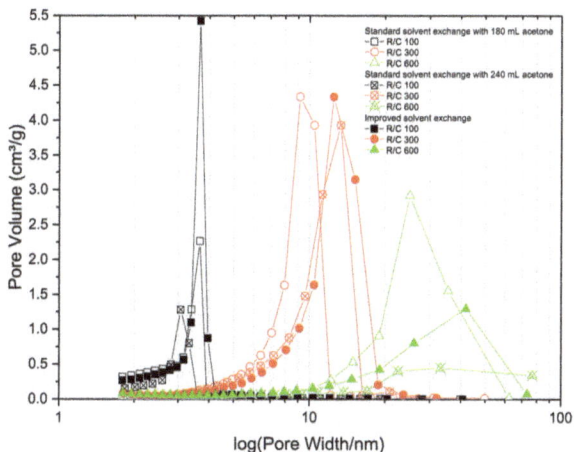

Figure 2. Effect of solvent exchange method on pore size distribution for resorcinol–formaldehyde xerogels with varied resorcinol:carbonate (R/C) molar ratio.

Table 1. Textural properties of resorcinol–formaldehyde xerogels prepared with standard and improved solvent exchange.

R/C	S_{BET} (m^2/g) Acetone Exchange Method			V_T (cm^3/g) Acetone Exchange Method			V_μ (cm^3/g) Acetone Exchange Method			$\bar{p}$ (nm) Acetone Exchange Method		
	Standard		Improved	Standard		Improved	Standard		Improved	Standard		Improved
	180 mL	240 mL		180 mL	240 mL		180 mL	240 mL		180 mL	240 mL	
100	480	470	580	0.33	0.33	0.46	0.052	0.046	0.059	3	3	3
200	470	530	500	0.54	0.71	0.73	0.040	0.056	0.052	5	5	6
300	420	430	470	0.78	0.93	1.05	0.043	0.052	0.060	8	10	11
400	370	300	220	0.97	0.95	0.99	0.046	0.035	0.033	12	14	24
500	300	220	230	0.97	0.96	1.17	0.039	0.033	0.034	16	24	29
600	230	110	220	1.01	0.44	0.81	0.036	0.019	0.036	24	27	22

S_{BET}—surface area from Brunauer-Emmett-Teller (BET) analysis; V_T—total pore volume determined from adsorption at p/p° ~1; V_μ—micropore volume determined using t-plot method; $\bar{p}$—average pore width from Barrett-Joyner-Halenda (BJH) analysis. Errors are omitted from the table as all values are reported to an accuracy less than the largest error for each variable.

2.2. Effect of Changing Solids Content

There is a tendency within the literature to use solids contents of ~20 $w/v\%$ in the production of RF gels [13,14,40,41]; however, the amount of solid material within the reaction volume would be expected to affect the solid:liquid ratio, hence, the final gel characteristics. Here, RF gels were synthesised over the range of solids content between 10 and 40 $w/v\%$, using R/C ratios of 100, 300 and 600. Note that these samples were prepared at 85 °C, using improved solvent exchange (see Section 2.1) and vacuum drying (see Section 2.3). For the samples synthesised using a solids content of 10 $w/v\%$, gelation was unsuccessful for R/C ratios greater than 600, hence, the range used in this study, but it should be noted that R/C ratio can be increased as the solids content increases but would not allow a direct comparison within this work, thus R/C 600 was the highest value studied here. For solids contents $\geq$20 $w/v\%$, some of the samples exhibited cracking during the drying stage, which affected their final characteristics.

From Table 2, it can be observed that, at constant R/C molar ratio, there is no significant change in specific surface area as mass content changes; however, the total pore volume is seen to increase with solids content, up to $\leq$30 $w/v\%$, after which point, the pore volume is slightly reduced at low R/C but still increases at higher R/C values. This can be ascribed to interplay between R/C ratio, i.e., particle nucleation number, and solids content, i.e., available material for particle growth; this means that the higher R/C ratios are more greatly affected by the additional mass available, due to the lower number of particles formed. The decrease at low R/C may be attributable to inhomogeneity during the gelation process, when no active agitation is applied, or possibly due to the increased mass per unit volume, which increases the relative density and reduces the void space available. Similarly, at constant R/C molar ratio, the average pore size increases with increasing solids content, again to 30 $w/v\%$, whereupon it decreases steadily with increasing reactant concentration. Increasing the mass of reactants at a fixed R/C ratio, increases both the monomer concentration and that of sodium carbonate, as the catalyst, which leads to an increase in the number of particles formed during gelation; this could result in the observed decrease in average pore size. It should be noted that the pore diameters determined for R/C 100 are constant at three nanometers; however, differentiation at this level is hindered by the size of the probe molecule, which only allows integer values to be reported.

Table 2. Textural properties of resorcinol–formaldehyde xerogels prepared using different percentage solids contents.

w/v% Solids	S_{BET} (m²/g) R/C Ratio			V_T (cm³/g) R/C Ratio			V_μ (cm³/g) R/C Ratio			$\bar{\varphi}$ (nm) R/C Ratio		
	100	300	600	100	300	600	100	300	600	100	300	600
10	500	370	-	0.36	0.85	-	0.057	0.037	-	3	9	-
20	500	490	280	0.32	0.91	1.00	0.065	0.064	0.046	3	8	18
25	550	410	190	0.42	1.00	1.07	0.054	0.042	0.030	3	10	32
30	570	490	260	0.46	1.08	1.17	0.055	0.064	0.045	3	9	28
35	570	450	260	0.45	0.98	1.23	0.051	0.050	0.038	3	9	27
40	540	550	330	0.44	1.07	1.53	0.048	0.077	0.056	3	9	29

S_{BET}—surface area from BET analysis; V_T—total pore volume determined from adsorption at p/p° ~1; V_μ—micropore volume determined using t-plot method; $\bar{\varphi}$—average pore width from BJH analysis. Errors are omitted from the table as all values are reported to an accuracy less than the largest error for each variable.

Figure 3 shows the pore size distribution of RF gel samples prepared at a constant R/C molar ratio of 300, and using different percentage solids contents. It can be seen that there is no significant change in the pore size distribution as the reactant concentration changes; however, it can be observed that RF gels with solids contents of 25 and 30 $w/v\%$ exhibit the narrowest distribution, with a sharp peak at ~15 nm. From Figure 4, meanwhile, it is obvious that altering the solids content has no major effect on overall shape of the adsorption–desorption isotherm of N_2, with all samples exhibiting Type IV isotherms [42]. The quantity of N_2 adsorbed increases with increasing relative pressure and a solids content of 30 $w/v\%$ shows the highest adsorption capacity of all levels tested. The combination of a discrete pore size distribution and high pore volume (Table 2) indicates that the selection of 20 $w/v\%$ in the synthetic matrix is in line with process optimization.

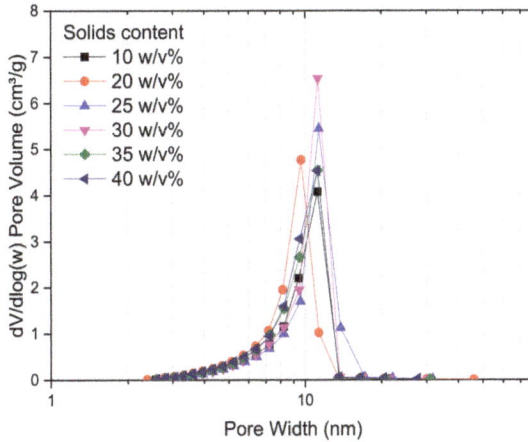

Figure 3. Pore size distribution obtained for resorcinol–formaldehyde xerogels synthesised using a resorcinol:cataylst molar ratio of 300 and varied percentage solids contents.

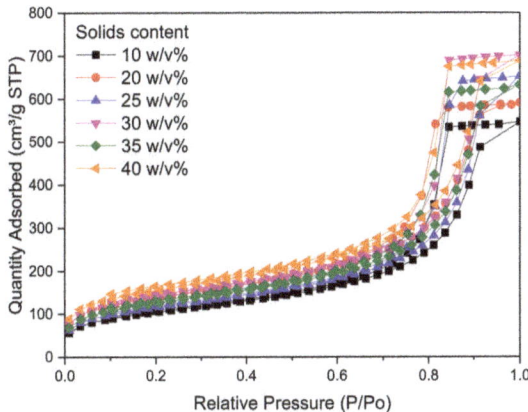

Figure 4. Nitrogen adsorption-desorption isotherms obtained for resorcinol–formaldehyde xerogels using a resorcinol:cataylst molar ratio of 300 and varied percentage solids content.

2.3. Ambient Pressure vs. Vacuum Drying

The final stage of gel preparation is drying of solvent exchanged gels, which, in this case, involves subcritical drying of the gels to remove acetone. The gels prepared in this way exhibit a higher degree of shrinkage; however, it is much easier to implement, and more economical, when compared to supercritical drying with CO_2. Usually, in order to make the drying process faster, and to ensure that the final materials have been dried thoroughly, vacuum drying is utilised. Maintaining a vacuum during drying is also not inexpensive, so it would be beneficial if RF gels could be dried under ambient pressure at elevated temperature, while retaining their final properties. Therefore, a series of gel samples were prepared, where the gel sample was divided in two halves post improved solvent exchange. This ensured that any effects observed within the final structure only resulted from the selected drying procedure. One half of the sample was dried for two days under vacuum at 85 °C, while the other half was dried under ambient pressure at 85 °C for one day and subsequently moved to the vacuum oven with the other sample half for one day of further drying, this time sub-atmospherically. Most of the drying process occurs during the first day; while the second day is used to remove the final traces of acetone remaining in the pores.

Table 3 shows the textural properties obtained for the gels prepared as outlined above. It can be observed that even though the gels dried under vacuum tend to have higher surface areas, pore volumes, micropore volumes, and larger average pore widths, the differences are insubstantial. This means that, if the requirements for the final material are not too strict, it should be possible to initially dry RF gels at ambient pressure, potentially even in the same oven as is used for gelation, since the temperatures are equivalent. From an industrial perspective, this could result in significant cost savings associated with the drying process of RF gels, and the handling of materials between unit operations, and could make such materials potentially cost-effective for new applications.

Table 3. Textural properties of resorcinol–formaldehyde gels dried at ambient pressure and under vacuum.

R/C	S_{BET} (m^2/g)		V_T (cm^3/g)		V_μ (cm^3/g)		$\bar{\varphi}$ (nm)	
	Drying Method		Drying Method		Drying Method		Drying Method	
	Ambient	Vacuum	Ambient	Vacuum	Ambient	Vacuum	Ambient	Vacuum
100	510	600	0.45	0.47	0.037	0.064	4	3
300	380	460	1.11	1.12	0.044	0.064	13	12
600	90	120	0.31	0.54	0.014	0.023	19	30

S_{BET}—surface area from BET analysis; V_T—total pore volume determined from adsorption at p/p° ~1; V_μ—micropore volume determined using t-plot method; $\bar{\varphi}$—average pore width from BJH analysis. Errors are omitted from the table as all values are reported to an accuracy less than the largest error for each variable.

2.4. Influence of Synthetic and Processing Temperature

In light of the three previous steps, it seems reasonable that the preparation of gels using 20 w/v% solids content, with an improved solvent exchange step and either ambient or vacuum drying should yield reasonably optimal materials. The constraint of several process variables also indicates that it should be possible to obtain materials with a high degree of reproducibility; however, this is dependent on control of one significant parameter, which can have significant impact on the overall process costs, i.e., temperature. The first stages of resorcinol–formaldehyde (RF) gel formation, immediately after mixing the components, are gelation and curing, which are usually carried out at elevated temperatures, and the final processing steps of gel production also include the use of a raised temperature during drying. Hence, the final parameter studied here was the influence of temperature within the manufacturing process. In all previous experiments, 85 °C was selected as the gelation and curing temperature as gels previously obtained at this temperature have exhibited a viable structure, and it is also a commonly used value in the literature, allowing further comparisons to be made to previously reported results [20,43,44]. It has, however, been shown that RF cluster particles begin to grow once the solution reaches a temperature of at least 55 °C [13], which indicates a minimum threshold for investigation; since water is used as the solvent, in the synthesis outlined above, the upper temperature limit is, therefore, set by the boiling point of water. Thus, the chosen temperature range studied was 45–95 °C, with 10 °C intervals. This allowed the region both above and below the temperature necessary for cluster growth to be probed to determine whether a viable gel structure can be established and maintained at temperatures approaching both (i) the boiling point of water and (ii) lower, less energy demanding, temperatures. R/C ratio was varied, as required, but all other synthesis parameters were kept constant as stated above; the only other change was that of oven temperature during the gelation and drying stages. Due to the enhanced performance observed above, improved solvent exchange was used exclusively, and the drying temperature, used in the vacuum stage, matched the gelation and curing temperatures, in order to restrict any post gelation changes in structure caused by exposure to a higher temperature during drying.

Table 4 shows the textural properties for gels synthesised at different temperatures, obtained from nitrogen adsorption analysis. Gels prepared at lower temperatures either did not gel or exhibited a very weak structure that did not withstand the drying process; this led to materials with a low degree of porosity or even to non-porous materials. The effect of temperature can be seen more clearly in Figure 5, where the influence of gel preparation temperature, and R/C ratio, on Brunauer-Emmett-Teller (BET) surface area is shown. It can be seen that, at low temperatures (45 and 55 °C), the surface areas obtained are very low, and are essentially independent of the R/C ratio used. At higher temperatures, the BET surface area seems to be only slightly dependent on temperature, and the effect of catalyst concentration dominates as the major factor in determining the final gel structure properties. These results are in disagreement with results from Tamon and Ishizaka [45] who reported that gelation temperature had no influence on the final gel structure. The difference is likely ascribed to the fact that their gelation step at either 25 or 50 °C was followed by a curing period of five days at 90 °C. Thus, the influence of the lower temperature gelation stage would have been masked by subsequent exposure to the same higher temperature during the curing stage.

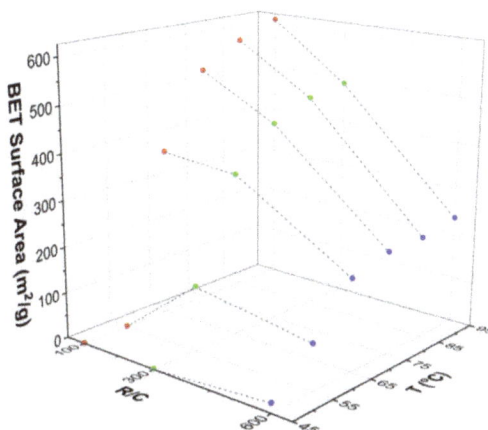

Figure 5. Dependence of BET surface area on resorcinol–formaldehyde xerogel preparation temperature and resorcinol:catalyst (R/C) molar ratio.

Table 4. Textural properties of resorcinol–formaldehyde xerogels prepared at different temperatures.

T (°C)	S_{BET} (m²/g)			V_T (cm³/g)			V_μ (cm³/g)			$\bar{\varphi}$ (nm)		
	R/C Ratio			R/C Ratio			R/C Ratio			R/C Ratio		
	100	300	600	100	300	600	100	300	600	100	300	600
45	-	<1	20	-	-	0.07	-	-	0.002	-	-	9
55	<1	140	100	-	0.14	0.48	-	0.010	0.011	-	4	22
65	370	350	200	0.22	0.52	0.82	0.054	0.036	0.027	3	6	20
75	530	440	220	0.37	0.77	0.82	0.064	0.052	0.030	3	8	21
85	580	470	220	0.46	1.05	0.81	0.059	0.060	0.036	3	11	22
95	610	490	230	0.52	1.18	0.92	0.057	0.064	0.038	4	12	24

S_{BET}—surface area from BET analysis; V_T—total pore volume determined from adsorption at p/p° ~1; V_μ—micropore volume determined using t-plot method; $\bar{\varphi}$—average pore width from BJH analysis. Errors are omitted from the table as all values are reported to an accuracy less than the largest error for each variable.

Pore size distributions for the suites of samples prepared using different temperatures, and R/C ratio 300, are presented in Figure 6, and the results show that the pore size distribution shifts towards larger pore diameters with increasing gelation temperature. This implies that gels prepared at higher temperatures develop stronger crosslinkages, which leads to a lower degree of shrinkage during the drying stage. It can also be observed that the total pore volume, which is given by the area under the pore size distribution curves, increases with increasing temperature, further supporting the theory that shrinkage is reduced within the stronger structures created at higher temperatures. The gels prepared at 45 °C exhibited such low porosity that the values are not even discernible in Figure 6, and are overlapped by other points; specific values are presented in Table 4.

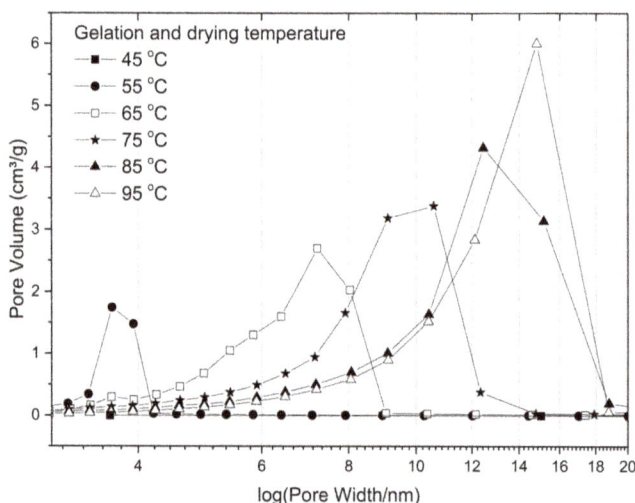

Figure 6. Effect of gelation temperature on pore size distributions for resorcinol–formaldehyde xerogels prepared using resorcinol:catalyst molar ratio of 300 and 20 w/v% solids content.

Morphological images of xerogel samples synthesised at 45 and 85 °C, with R/C ratios 100 and 600, are shown in Figure 7. It can be observed that the samples prepared with R/C ratio 100 do not show any significant textural features at this macroscopic level, which is expected considering the results from nitrogen sorption measurements. The pore size for these samples is below the limit at this magnification and due to the porous nature of the samples, it was not possible to achieve higher magnifications without using a higher thickness of gold coating, which would obscure any fine textural features. By contrast, there is a clear difference in morphology between the samples prepared with R/C 600 at different temperatures. The xerogel prepared at 85 °C (Figure 7d) exhibits a typical porous structure, composed of RF clusters crosslinked into a 3D network with some of the macropores clearly visible. While there are visible differences between samples prepared at 85 °C (Figure 7b,d), the xerogels prepared at 45 °C (Figure 7a,c) exhibit a very similar structure independent of catalyst amount. This agrees with the textural data obtained from nitrogen sorption measurements.

It is evident from these results that, in order to obtain a viable gel structure capable of enduring the drying process, the gelation temperature must be in excess of 55 °C, as suggested by Taylor et al. [13]; however, increasing the temperature further does not seem to have a significant impact on the surface area obtained. The other textural variables are affected slightly and it may be required to use elevated temperatures to optimise a particular variable or enhance the crosslinking within the final gel. This information could be used in process optimization of RF gel manufacture to reduce the heating costs associated with the gelation and drying processes for a specific set of required textural characteristics, as defined by a selected application.

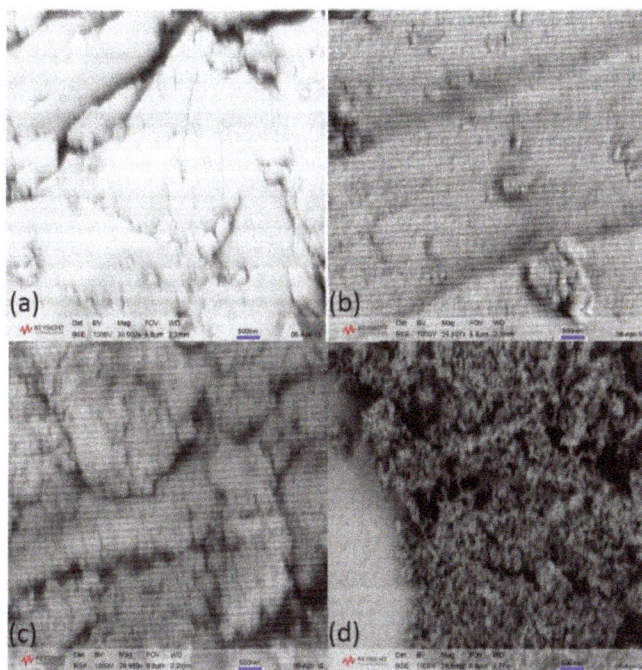

Figure 7. SEM micrographs of resorcinol–formaldehyde xerogels prepared at (**a**) 45 °C with resorcinol:catalyst molar ratio of 100, (**b**) 85 °C with resorcinol:catalyst molar ratio of 100, (**c**) 45 °C with resorcinol:catalyst molar ratio of 600, and (**d**) 85 °C with resorcinol:catalyst molar ratio of 600 at 30,000× magnification.

3. Conclusions

The work presented here demonstrates the need to carefully control the synthesis and process parameters used in RF gel production, in order to obtain the optimal material for a given application. Solids content is integral to gel viability, with low solids contents resulting in weaker structures that fail to gel at higher R/C ratios, and very high solids contents resulting in increased densification of the material and a reduction in porosity. It was observed that 30 w/v% represents an upper bound for solids content in the systems studied here, and such materials exhibited the highest accessible pore volume; however, surface area was unaffected by increased mass, at constant R/C. It is suggested that the increased mass of reactants (both monomer and catalyst) increased particle number and decreased average pore size. Within the systems studied, those gels created using solids contents of 20–30 w/v% exhibited the narrowest distribution; thus, the combination of discrete pore size distribution and high pore volume, with lower reactant costs, indicates 20 w/v% is optimal for gel production. In line with previous studies, a minimum temperature of 55 °C was shown to be critical in viable gel formation; gels prepared at lower temperatures either did not gel or exhibited a very weak structure with low or negligible porosity, independent of R/C. Gels prepared at higher temperatures showed insignificant changes in surface area with temperature, with the effect of catalyst concentration dominating gel formation; while pore diameter increases with increasing gelation temperature, due to stronger crosslinkages, hence, a lower degree of shrinkage during processing. This indicates that, while the gelation temperature must be in excess of 55 °C, increasing the temperature further has little impact on the final surface area, allowing a lower temperature to be used for gel synthesis if this is a key measure of gel performance. Post-synthesis, the regular replacement of the solvent exchange fluid has a marked positive effect on total pore volume, leading to pores with larger average diameters, which is ascribed to a reduction in shrinkage during the drying stage, due to the increased driving force for water removal, hence, lower stresses being exerted on the pore walls during processing. It is, therefore, not necessary to increase the amount of solvent used within the exchange but it is imperative to increase the number of solvent changes over the exchange period. Finally, the differences between gels dried at atmospheric and sub-atmospheric pressure show little difference in their textural character, hence, it may be possible to dry RF gels at ambient pressure, potentially even in the same oven as gelation, to reduce both heating and pump costs. Combined, these results provide guidance to reduce the costs of RF gel manufacture, without impinging on the desired qualities of the materials produced.

4. Materials and Methods

4.1. Sample Preparation

Unless otherwise stated, all resorcinol–formaldehyde (RF) gel samples were prepared using an analogous procedure, excepting for the specific parameter investigated in each section of the study. All chemicals were used as received from the supplying company, and deionised water was produced in-house (Millipore Elix® 5 with Progard® 2 (Merck, Watford, UK)). Firstly, the appropriate amount of resorcinol (Sigma Aldrich, Gillingham, UK, ReagentPlus, 99%) was added to a premeasured volume of deionized water in a jar containing a magnetic stirrer bar. Upon dissolution of all of the added resorcinol, a corresponding amount of sodium carbonate (Sigma Aldrich, anhydrous, ≥99.5%), on a molar basis, was weighed out and added to the solution. As outlined above, sodium carbonate acts as a catalyst, by a combination of increasing the solution pH in the basic region via hydrolysis of the carbonate ion, and by the introduction of sodium ions, which, it has been suggested, assist in the addition of formaldehyde to resorcinol [46]. Catalyst concentration is expressed as resorcinol/catalyst molar ratio (R/C) and the range studied here is R/C 100–600. After all solids were dissolved, the required amount of formaldehyde, in the form of formalin solution (Sigma Aldrich, 37 wt % formaldehyde in water, containing 10–15 wt % methanol as a polymerization inhibitor), was added, and the solution was stirred in a closed jar for 30 min. All samples were prepared with 20 *w/v*% solids content, unless otherwise stated, and the total volume used was 60 mL, made up of water and methanol, contributed by the formalin solution used. At the end of the period of agitation, stirrer bars were removed from the solution, and the jar lid was hand-tightened, before moving the jar to an oven (Memmert UFE400, Schwabach, Germany) preheated to 85 °C, unless otherwise stated. Samples formed during this study gelled within 1–2 h [13]; however, samples were left to cure for three days in order to ensure sufficient time for crosslinking to occur. After three days, the jars containing the gels were removed from the oven and left to cool to room temperature. The formed gels were cut into smaller pieces before washing and solvent exchange with acetone (Sigma Aldrich, ≥99.5%). Standard solvent exchange involved addition of ~180 or ~240 mL of acetone to the drained gel, before resealing the lid and, in order to minimise acetone losses, wrapping with paraffin film. Sealed jars were put on a shaker unit (VWR 3500 Analog Orbital Shaker, Lutterworth, UK) and agitated for three days. In the improved solvent exchange method, the exchanged acetone was drained and replaced with 80 mL of fresh solvent on each successive day for three days. After three days of either solvent exchange method, the gel was drained and placed in a vacuum oven (Townson and Mercer 1425 Digital Vacuum Oven, Stretford, UK), preheated to 85 °C (or, in the case of the temperature study samples, the drying temperature was set to match the curing temperature), to dry for two days. Finally, the sample was transferred to a labelled sample tube for storage.

4.2. Sample Characterisation

Nitrogen adsorption-desorption measurements were used to obtain textural properties for the RF gel samples prepared in this study. Nitrogen sorption was performed at −196 °C using a Micromeritics ASAP 2420 (Hexton, UK) surface area and porosity analyser. Prior to analysis, samples were outgassed under vacuum below 10 µmHg at 50 °C for 30 min and then at 110 °C for 2 h; except for samples where the influence of temperature was investigated, for these samples, outgassing temperatures matched the gelation and drying temperatures used, and the time for outgassing was adjusted accordingly to ensure removal of all volatile contaminant species. Samples were analysed using a 40 pressure point adsorption and 30 pressure point desorption cycle. All samples were characterised for surface area (m^2/g), using Brunauer-Emmett-Teller (BET) theory [47], and the Rouquerol correction for microporous samples [42]; total pore volume (cm^3/g); micropore volume (cm^3/g) from the t-plot method [48]; and average pore size (nm) from the Barrett-Joyner-Halenda method [49].

Scanning electron microscopy (SEM) micrographs were recorded in backscattered mode at 1000 V using a Field Emission Scanning Electron Microscope (Keysight, U9320B, Wokingham, UK) at magnification 30,000×. Prior to analysis, samples were ground into a fine powder, coated with a 10 nm gold layer using an EM ACE 200 sputter-coater (Leica Inc., Milton Keynes, UK), and attached to aluminium stubs with carbon tape.

Acknowledgments: Martin Prostredný thanks the University of Strathclyde and the Department of Chemical and Process Engineering for financial support. Mohammed G. M. Abduljalil thanks the Libyan Government for financial support. The authors would also like to acknowledge that the SEM analysis was carried out in the CMAC (Continuous Manufacturing and Crystallisation) National Facility, housed within the University of Strathclyde's Technology and Innovation Centre, and funded with a UKRPIF (UK Research Partnership Institute Fund) capital award, SFC (Scottish Funding Council) ref. H13054, from the Higher Education Funding Council for England (HEFCE).

Author Contributions: Martin Prostredný, Mohammed G. M. Abduljalil, Paul A. Mulheran and Ashleigh J. Fletcher contributed equally to conceiving and designing the experiments, analysing the data and writing and revising the manuscript. Martin Prostredný and Mohammed G. M. Abduljalil performed the experiments.

Conflicts of Interest: The authors declare no conflict of interest.

References

1. Pekala, R.W. Organic aerogels from the polycondensation of resorcinol with formaldehyde. *J. Mater. Sci.* **1989**, *24*, 3221–3227. [CrossRef]
2. Marie, J.; Berthon-Fabry, S.; Chatenet, M.; Chaînet, E.; Pirard, R.; Cornet, N.; Achard, P. Platinum supported on resorcinol–formaldehyde based carbon aerogels for PEMFC electrodes: Influence of the carbon support on electrocatalytic properties. *J. Appl. Electrochem.* **2007**, *37*, 147–153. [CrossRef]
3. Job, N.; Marie, J.; Lambert, S.; Berthon-Fabry, S.; Achard, P. Carbon xerogels as catalyst supports for PEM fuel cell cathode. *Energy Convers. Manag.* **2008**, *49*, 2461–2470. [CrossRef]
4. Smirnova, A.; Dong, X.; Hara, H.; Vasiliev, A.; Sammes, N. Novel carbon aerogel-supported catalysts for PEM fuel cell application. *Int. J. Hydrogen Energy* **2005**, *30*, 149–158. [CrossRef]
5. Robertson, C.; Mokaya, R. Microporous activated carbon aerogels via a simple subcritical drying route for CO_2 capture and hydrogen storage. *Microporous Mesoporous Mater.* **2013**, *179*, 151–156. [CrossRef]
6. Gross, A.F.; Vajo, J.J.; Van Atta, S.L.; Olson, G.L. Enhanced hydrogen storage kinetics of $LiBH_4$ in nanoporous carbon scaffolds. *J. Phys. Chem. C* **2008**, *112*, 5651–5657. [CrossRef]
7. Yamamoto, T.; Endo, A.; Ohmori, T.; Nakaiwa, M. Porous properties of carbon gel microspheres as adsorbents for gas separation. *Carbon* **2004**, *42*, 1671–1676. [CrossRef]
8. Dong, Y.-R.; Nakao, M.; Nishiyama, N.; Egashira, Y.; Ueyama, K. Gas permeation and pervaporation of water/alcohols through the microporous carbon membranes prepared from resorcinol/formaldehyde/quaternary ammonium compounds. *Sep. Purif. Technol.* **2010**, *73*, 2–7. [CrossRef]
9. Alcántara, R.; Lavela, P.; Ortiz, G.F.; Tirado, J.L. Carbon microspheres obtained from resorcinol-formaldehyde as high-capacity electrodes for sodium-ion batteries. *Electrochem. Solid-State Lett.* **2005**, *8*, A222–A225. [CrossRef]
10. Glora, M.; Wiener, M.; Petricevic, R.; Probstle, H.; Fricke, J. Integration of carbon aerogels in PEM fuel cells. *J. Non-Cryst. Solids* **2001**, *285*, 283–287. [CrossRef]
11. Lu, X.; Arduini-Schuster, M.; Kuhn, J.; Nilsson, O.; Fricke, J.; Pekala, R. Thermal conductivity of monolithic organic aerogels. *Science* **1992**, *255*, 971–972. [CrossRef] [PubMed]

12. Feng, J.; Zhang, C.; Feng, J. Carbon fiber reinforced carbon aerogel composites for thermal insulation prepared by soft reinforcement. *Mater. Lett.* **2012**, *67*, 266–268. [CrossRef]
13. Taylor, S.J.; Haw, M.D.; Sefcik, J.; Fletcher, A.J. Gelation mechanism of resorcinol-formaldehyde gels investigated by dynamic light scattering. *Langmuir* **2014**, *30*, 10231–10240. [CrossRef] [PubMed]
14. Taylor, S.J.; Haw, M.D.; Sefcik, J.; Fletcher, A.J. Effects of secondary metal carbonate addition on the porous character of resorcinol–formaldehyde xerogels. *Langmuir* **2015**, *31*, 13571–13580. [CrossRef] [PubMed]
15. Rey-Raap, N.; Angel Menéndez, J.; Arenillas, A. Simultaneous adjustment of the main chemical variables to fine-tune the porosity of carbon xerogels. *Carbon* **2014**, *78*, 490–499. [CrossRef]
16. Rey-Raap, N.; Angel Menéndez, J.; Arenillas, A. RF xerogels with tailored porosity over the entire nanoscale. *Microporous Mesoporous Mater.* **2014**, *195*, 266–275. [CrossRef]
17. Job, N.; Gommes, C.J.; Pirard, R.; Pirard, J.-P. Effect of the counter-ion of the basification agent on the pore texture of organic and carbon xerogels. *J. Non-Cryst. Solids* **2008**, *354*, 4698–4701. [CrossRef]
18. Fairen-Jimenez, D.; Carrasco-Marin, F.; Moreno-Castilla, C. Porosity and surface area of monolithic carbon aerogels prepared using alkaline carbonates and organic acids as polymerization catalysts. *Carbon* **2006**, *44*, 2301–2307. [CrossRef]
19. Laskowski, J.; Milow, B.; Ratke, L. Subcritically dried resorcinol-formaldehyde aerogels from a base-acid catalyzed synthesis route. *Microporous Mesoporous Mater.* **2014**, *197*, 308–315. [CrossRef]
20. Al-Muhtaseb, S.A.; Ritter, J.A. Preparation and properties of resorcinol–formaldehyde organic and carbon gels. *J. Adv. Mater.* **2003**, *15*, 101–114. [CrossRef]
21. El Khatat, A.M.; Al-Muhtaseb, S.A. Advances in tailoring resorcinol-formaldehyde organic and carbon gels. *Adv. Mater.* **2011**, *23*, 2887–2903. [CrossRef] [PubMed]
22. Tamon, H.; Ishizaka, H.; Mikami, M.; Okazaki, M. Porous structure of organic and carbon aerogels synthesized by sol-gel polycondensation of resorcinol with formaldehyde. *Carbon* **1997**, *35*, 791–796. [CrossRef]
23. Gaca, K.Z.; Parkinson, J.A.; Sefcik, J. Kinetics of early stages of resorcinol-formaldehyde polymerization investigated by solution-phase nuclear magnetic resonance spectroscopy. *Polymer* **2017**, *110*, 62–73. [CrossRef]
24. Rojas-Herrera, J.; Lozano, P.C. Mitigation of anomalous expansion of carbon xerogels and controllability of mean-pore-size by changes in mold geometry. *J. Non-Cryst. Solids* **2017**, *458*, 22–27. [CrossRef]
25. Pierre, A.C.; Pajonk, G.M. Chemistry of aerogels and their application. *Chem. Rev.* **2002**, *102*, 4243–4265. [CrossRef] [PubMed]
26. Lide, D.R. *CRC Handbook of Chemistry and Physics, Internet Version 2006*; Taylor and Francis: Boca Raton, FL, USA, 2006.
27. Barbieri, O.; Ehrburger-Dolle, F.; Rieker, T.P.; Pajonk, G.M.; Pinto, N.; Rao, A.V. Small-angle X-ray scattering of a new series of organic aerogels. In Proceedings of the 6th International Symposium on Aerogels (ISA-6), Albuquerque, NM, USA, 8–11 October 2000; pp. 109–115.
28. Fricke, J.; Tillotson, T. Aerogels: Production, characterization, and applications. *Thin Solid Films* **1997**, *297*, 212–223. [CrossRef]
29. Pekala, R.W.; Kong, F.M. A synthetic route to organic aerogels-mechanism, structure, and properties. *J. Phys. Colloq.* **1989**, *50*, 33–40. [CrossRef]
30. Job, N.; Pirard, R.; Marien, J.; Pirard, J.P. Porous carbon xerogels with texture tailored by pH control during sol-gel process. *Carbon* **2004**, *42*, 619–628. [CrossRef]
31. Tamon, H.; Ishizaka, H.; Yamamoto, T.; Suzuki, T. Preparation of mesoporous carbon by freeze drying. *Carbon* **1999**, *37*, 2049–2055. [CrossRef]
32. Czakkel, O.; Marthi, K.; Geissler, E.; Laszlo, K. Influence of drying on the morphology of resorcinol-formaldehyde-based carbon gels. *Microporous Mesoporous Mater.* **2005**, *86*, 124–133. [CrossRef]
33. Yamamoto, T.; Nishimura, T.; Suzuki, T.; Tamon, H. Control of mesoporosity of carbon gels prepared by sol-gel polycondensation and freeze drying. *J. Non-Cryst. Solids* **2001**, *288*, 46–55. [CrossRef]
34. Job, N.; Thery, A.; Pirard, R.; Marien, J.; Kocon, L.; Rouzaud, J.N.; Beguin, F.; Pirard, J.P. Carbon aerogels, cryogels and xerogels: Influence of the drying method on the textural properties of porous carbon materials. *Carbon* **2005**, *43*, 2481–2494. [CrossRef]
35. Jabeen, N.; Mardan, A. Effect of water removal on the textural properties of resorcinol/formaldehyde gels by azeotropic distillation. *J. Mater. Sci.* **1998**, *33*, 5451–5453. [CrossRef]

36. Schwertfeger, F.; Frank, D.; Schmidt, M. Hydrophobic waterglass based aerogels without solvent exchange or supercritical drying. *J. Non-Cryst. Solids* **1998**, *225*, 24–29. [CrossRef]

37. Vargaftik, N.; Volkov, B.; Voljak, L. International tables of the surface tension of water. *J. Phys. Chem. Ref. Data* **1983**, *12*, 817–820. [CrossRef]

38. Mukai, S.R.; Tamitsuji, C.; Nishihara, H.; Tamon, H. Preparation of mesoporous carbon gels from an inexpensive combination of phenol and formaldehyde. *Carbon* **2005**, *43*, 2628–2630. [CrossRef]

39. Jasper, J.J. The surface tension of pure liquid compounds. *J. Phys. Chem. Ref. Data* **1972**, *1*, 841–1010. [CrossRef]

40. Zanto, E.J.; Al-Muhtaseb, S.A.; Ritter, J.A. Sol-gel-derived carbon aerogels and xerogels: Design of experiments approach to materials synthesis. *Ind. Eng. Chem. Res.* **2002**, *41*, 3151–3162. [CrossRef]

41. Berthon, S.; Barbieri, O.; Ehrburger-Dolle, F.; Geissler, E.; Achard, P.; Bley, F.; Hecht, A.-M.; Livet, F.; Pajonk, G.M.; Pinto, N. DLS and SAXS investigations of organic gels and aerogels. *J. Non-Cryst. Solids* **2001**, *285*, 154–161. [CrossRef]

42. Thommes, M.; Kaneko, K.; Neimark, A.V.; Olivier, J.P.; Rodriguez-Reinoso, F.; Rouquerol, J.; Sing, K.S.W. Physisorption of gases, with special reference to the evaluation of surface area and pore size distribution (iupac technical report). *Pure Appl. Chem.* **2015**, *87*, 1051–1069. [CrossRef]

43. Calvo, E.; Menéndez, J.; Arenillas, A. Influence of alkaline compounds on the porosity of resorcinol-formaldehyde xerogels. *J. Non-Cryst. Solids* **2016**, *452*, 286–290. [CrossRef]

44. Lin, C.; Ritter, J.A. Effect of synthesis pH on the structure of carbon xerogels. *Carbon* **1997**, *35*, 1271–1278. [CrossRef]

45. Tamon, H.; Ishizaka, H. Influence of gelation temperature and catalysts on the mesoporous structure of resorcinol-formaldehyde aerogels. *J. Colloid Interface Sci.* **2000**, *223*, 305–307. [CrossRef] [PubMed]

46. GrenierLoustalot, M.F.; Larroque, S.; Grande, D.; Grenier, P.; Bedel, D. Phenolic resins: 2. Influence of catalyst type on reaction mechanisms and kinetics. *Polymer* **1996**, *37*, 1363–1369. [CrossRef]

47. Brunauer, S.; Emmett, P.H.; Teller, E. Adsorption of gases in multimolecular layers. *J. Am. Chem. Soc.* **1938**, *60*, 309–319. [CrossRef]

48. Lowell, S.; Shields, J.E.; Thomas, M.A.; Thommes, M. Micropore analysis. In *Characterization of Porous Solids and Powders: Surface Area, Pore Size and Density*; Springer: Berlin, Germany, 2004; pp. 129–156.

49. Barrett, E.P.; Joyner, L.G.; Halenda, P.P. The determination of pore volume and area distributions in porous substances. I. Computations from nitrogen isotherms. *J. Am. Chem. Soc.* **1951**, *73*, 373–380. [CrossRef]

gels

MDPI

Communication

A Simple Method of Interpretating the Effects of Electric Charges on the Volume Phase Transition of Thermosensitive Gels

Hiroshi Maeda [1],*, Shigeo Sasaki [2], Hideya Kawasaki [3] and Rie Kakehashi [4]

[1]   Professor Emeritus, Kyushu University, 46-6 Kashii 3-chome, Higashi-ku, Fukuoka 813-0011, Japan
[2]   Department of Chemistry, Faculty of Sciences, Kyushu University, Motooka 744, Nishi-ku, Fukuoka 819-0395, Japan
[3]   Faculty of Chemistry, Materials and Bioengineering, Kansai University, 3-3-35 Yamate-cho, Suita 564-8680, Japan; hkawa@kansai-u.ac.jp
[4]   Osaka Research Institute of Industrial Science and Technology, 1-6-50 Morinomiya, Joto-ku, Osaka 536-8553, Japan; rie@omtri.or.jp
*   Correspondence: maeda@chem.kyushu-univ.jp; Tel.: +81-92-681-8080

Received: 22 February 2018; Accepted: 13 March 2018; Published: 19 March 2018

Abstract: Various apparently inconsistent experimental observations have been reported on the effects of electric charges on the volume phase transition of thermosensitive gels. A simple method of interpretating these results is presented.

Keywords: volume phase transition; effects of electric charge; swelling of thermosensitive gels

1. Introduction

As the solvent in a polymer solution becomes poor, a phase separation will take place resulting in a phase equilibrium between two solutions of high and low polymer concentrations. What will happen, however, when the polymer chains are connected to form a network? This interesting question was first noticed and considered by Dušek and the possibility of phase transitions of networks has been discussed [1]. The volume phase transition (a discontinuous volume change) of gels, predicted by Dušek, was first experimentally observed by Tanaka on polyacrylamide gels in water-acetone mixtures [2]. The observed transition-like volume change was later ascribed to the effects of electric charges originating from acrylic acid groups as a result of hydrolysis of acrylamide groups. Later, Ilavsky et al. demonstrated [3,4], using copolymers (NDEA-*co*-MA) of *N*,*N*-diethylacrylamide (NDEA) and methacrylic acid (MA), that volume changes of non-ionic NDEA gels are continuous, while ionic (NDEA-*co*-MA) copolymer gels exhibited discrete volume changes. In this way, the idea was once prevailing that discontinuous volume changes of gels are favored by electric charges incorporated in polymer chains. In 1984, Hirokawa et al. observed the volume phase transition in a non-ionic thermosensitive poly(*N*-isopropylacrylamide) (NIPA) gels [5]. Then, effects of electric charges on the volume change of gels were examined by Hirotsu et al. [6] using copolymer gels (NIPA-*co*-AA) of NIPA and acrylic acid (AA). Their results indicated that introduction of electric charges enhanced the transition-like behavior observed for non-ionic NIPA gels. On the other hand, a study on another NIPA-copolymer gel, copolymers of NIPA and methacrylamidopropyltrimethylammonium chloride (MAPTAC), has indicated that transition-like volume changes of non-ionic NIPA gels disappear for copolymers containing MAPTAC more than 2 mol % and the volume changes of the latter gels are continuous [7]. These inconsistent results obtained on two NIPA-*co*-ionic component gels, NIPA-*co*-AA gels and NIPA-*co*-MAPTAC gels, have been puzzling. One way to reconcile the inconsistent results is a notion that AA and MAPTAC are different chemical entities and, hence, polymer-solvent and polymer-polymer interactions may differ significantly for the two NIPA copolymer gels [7]. Later, copolymer gels of NIPA with two kinds of ionic groups, one carboxyl group and the other amino group, have exhibited continuous temperature-induced volume changes as a result of the ionization of either ionic group, indicating chemically different groups have little consequences in this example [8]. Meanwhile, it has been concluded that NIPA-*co*-AA gels exhibit continuous volume changes with ionization [9,10], which is a similar behavior as NIPA-*co*-MAPTAC gels [7], but contrary to the reported behavior on the same kind of gel [6]. Thus, the results on two NIPA copolymer gels clearly indicate that introduction of electric charges changes a discontinuous volume change of NIPA gels to a continuous one [7–10]. The conclusion is further confirmed on another copolymer gel, NIPA-*co*-styrene sulfonate (SS) gel [9].

Summing up the experimental results described above, it is highly likely that an important mechanism is present regarding the effects of electric charges on the temperature-induced volume changes of gels, other than arising from different chemical properties of ionic groups. In this Communication we wish to propose a simple model to account for the charge effect under a highly approximate level of analysis.

It is pertinent to state here that continuous volume changes have been observed even for non-ionic NIPA gels, which suggests the volume phase transition of NIPA gels locates rather near the critical point. Additionally, we discuss only the case of thermosensitive non-ionic gels and the volume phase transition of ionic gels, in general, is completely outside the scope of this Communication.

2. The Swelling Equations of Gels

The chemical potential of solvent water μ_w is measured from the reference state of pure water $\mu_w{}^0$, $\Delta\mu_w = \mu_w - \mu_w{}^0$. Then, $\Delta\mu_w$ can be written as Equation (1) in terms of temperature T, and the gas constant R, and the number of moles of the solvent water n_1:

$$\Delta\mu_w/RT = \ln(1 - \phi) + \phi + \chi\phi^2 + [\partial G_{elas}/\partial n_1 + \Delta\mu_w{}^{ion}]/RT, \tag{1}$$

Here, m G_{elas} represents the elastic free energy and $\Delta\mu_w{}^{ion}$ represents the contribution arising from the introduction of ionic groups together with associated counterions. The volume fraction of polymer and the interaction parameter of polymer with solvent are denoted as ϕ and χ, respectively. We write the gel volume as V, that in the dry state as V_d, and hence $\phi = V_d/V$. The gel volume V_0 corresponds to the state in which polymer chains are in the unperturbed state.

2.1. The Elastic Free Energy of Gels G_{elas}

We denote the number of moles of active polymer chains in the whole gel as v and v^* as v/V_d. We assume the relation $\phi_0 V_0 = \phi V = \phi_d \cdot V_d$, where ϕ_d and ϕ_0 correspond, respectively, to the polymer volume fractions in the dry gel and in the unperturbed state gel. The linear deformation of the network α is given as follows:

$$\alpha = (V/V_0)^{1/3} = \phi_0{}^{1/3} \cdot \phi^{-1/3}, \tag{2}$$

For the affine network with functionality = 4, G_{elas} is given as follows [11]:

$$G_{elas} = (3vRT/2)\,(\alpha^2 - 1 - \ln\alpha), \tag{3}$$

and:

$$\partial G_{elas}/\partial n_1 = 3vRT\,[\alpha - 1/(2\alpha)]\,(\partial\alpha/\partial n_1) = v\cdot(V_1/V_0)\cdot RT\,[(\phi/\phi_0)^{1/3} - (1/2)\,(\phi/\phi_0)], \tag{4}$$

Here, the partial molar volume of solvent water is denoted as V_1.

2.2. Uncharged Gels ($\Delta\mu_w{}^{ion} = 0$)

From Equations (1) and (4), we have the following expressions for $\Delta\mu_w/RT$:

$$\Delta\mu_w/RT = \ln(1-\phi) + \phi + \chi\phi^2 + v(V_1/V_0)\cdot[(\phi/\phi_0)^{1/3} - (1/2)\cdot(\phi/\phi_0)], \tag{5}$$

$$= \ln(1-\phi) + \phi + \chi\phi^2 + v^*\cdot RT[(V_1/\phi_d)]\cdot[\phi_d{}^{2/3} <\alpha^2>_0 \phi^{1/3} - (1/2)\cdot\phi], \tag{5'}$$

Here, the isotropic deformation factor $<\alpha^2>_0$ is defined as $<\alpha^2>_0 = (V_d/V_0)^{2/3} = (\phi_0/\phi_d)^{2/3}$. There is a positive maximum of $\Delta\mu_w/RT$ in the region of very small ϕ, due to the situation that polymer chains connected through crosslinks cannot be diluted indefinitely. The equilibrium swelling volume V is determined by the condition of $\Delta\mu_w = 0$. When there are two maxima and one minimum of $\Delta\mu_w/RT$ satisfying the conditions $(\Delta\mu_w/RT)_{min} < 0$ and $(\Delta\mu_w/RT)_{max} > 0$, the gel volume change induced by a change in χ can be discontinuous (transition-like behavior). The critical end point for the volume phase transition is given as follows:

$$\partial(\Delta\mu_w/RT)/\partial\phi = 0 \text{ and } \partial^2(\Delta\mu_w/RT)/(\partial\phi)^2 = 0, \tag{6}$$

Dušek and Patterson have examined the critical condition in terms of Equation (5') under the assumption that $\phi_d = 1$ and χ is a constant independent of ϕ [1]. They concluded that there is no unique set of parameter values [ϕ, χ, v^*, $<\alpha^2>_0$] corresponding to the critical point.

On the other hand, increase of χ values with increasing polymer concentration has been frequently observed for solutions of linear polymers in poor solvents [11]. Erman and Flory have shown that discontinuous volume change of gels is possible if relevant ϕ-dependent χ parameters are taken into account [11]. Below, we assume χ parameters that linearly depend on ϕ:

$$\chi = \chi_1 + \chi_2 \phi, \tag{7}$$

3. Effects of Electric Charge on the Swelling Behavior of Gels

At least, the following three effects are noted:

1. Contribution from the mixing with counterions.
2. Effects on G_{elas} (introduction of electric charges generally makes polymer chains stiffer).
3. Introduction of ionic groups is nothing more than the introduction of different chemical species into polymer chains. Accordingly, two (or more) kinds of χ parameters are required to describe the polymer-solvent interactions. When a kind of average χ parameter is introduced, it may significantly differ from that for the parent non-ionic polymer.

In the present highly approximate analyses, two possible contributions 2 and 3 above are ignored and only the counterion contribution is considered. Two frequently-observed phenomena with the introduction of electric charges are: (a) the concentrated phase (collapsed phase) becomes more unstable, which corresponds to an increase in the transition temperature in the case of NIPA gels; and (b) the increased swollen gel volume results in an enlarged volume change at the transition.

Osmotic Pressure of Small Ions

The total mole number of counterions in the gel is iv for the case of no added salt, where i denotes the number of charges per chain. Then, $\Delta\mu_w{}^{ion}/RT$ is given as follows in terms of the osmotic coefficient g:

$$\Delta\mu_w{}^{ion}/RT = -gV_1 (i\cdot v/V) = -g\cdot i\cdot(v\cdot V_1/V_0)(\phi/\phi_0), \tag{8}$$

It is to be noted that there are no small ions in the water outside the gel if no salt is added.

In the case of gels immersed in solutions containing a uni-univalent salt of the same counterion species, the Donnan osmotic pressure is approximately evaluated as follows by assuming the additivity rule [12] and the ideal behavior of the salt ions:

$$\Delta\mu_w{}^{ion}/RT = -g \cdot i \cdot (v \cdot V_1/V_0) (\phi/\phi_0) - 2 V_1 (Cs - Cs'), \tag{9}$$

As evident in Equations (8) or (9), $\Delta\mu_w{}^{ion}$ is negative as a result of the introduction of ionic groups. The negative contribution to $\Delta\mu_w{}^{ion}$ arises not from the electric interactions among charges, but from the introduction of counterions. Combining Equations (5) and (9), we have the following final expression describing the swelling of thermosensitive ionic gels:

$$\Delta\mu_w/RT = \ln(1 - \phi) + \phi + \chi\phi^2 + v \cdot (V_1/V_0) [(\phi/\phi_0)^{1/3} - (1/2) \cdot (\phi/\phi_0)] - g \, i \cdot (v \cdot V_1/V_0) (\phi/\phi_0) - 2V_1 \cdot (Cs - Cs'), \tag{10}$$

As described in the introduction part, various and inconsistent experimental results have been reported in relation to the effects of electric charges on the volume change of thermosensitive gels. Different behaviors with the introduction of electric charges will be classified into three groups: (a) a transition-like volume change is induced [3,4]; (b) the transition–like behavior observed on the uncharged gel is preserved [6]; or (c) the transition–like behavior observed on the uncharged gel disappears [7–10], but the transition-like behavior is restored on the addition of a salt [7,9].

A simple way of understanding these diverse behaviors (a)–(c) above will be presented below. To simplify the discussion, we calculate the swelling behaviors in terms of Equation (11), a simplified version of Equation (10), with $g = \phi_0 = 1$. In the absence of salt ($Cs = Cs' = 0$):

$$\Delta\mu_w/RT = \ln(1 - \phi) + \phi + \chi\phi^2 + v(V_1/V_0) (\phi^{1/3} - \phi/2) - i \cdot (v \cdot V_1/V_0) \cdot \phi. \tag{11}$$

A situation corresponding to case (a) above may be indicated in Figure 1. Curve 1 in the figure represents a possible swelling curve of the uncharged gel, calculated with Equation (11) with $i = 0$. When the contribution from $\Delta\mu_w{}^{ion}/RT$, the last term in Equation (11), represented with a straight line 2 is added, curve 3 results which indicates the transition-like swelling behavior.

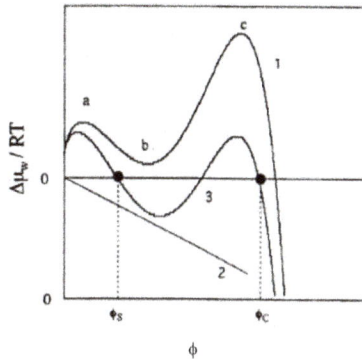

Figure 1. Calculated swelling curves in terms of Equation (11) in the case of the charge-induced volume phase transition. ($v \, V_1/V_0$) = 0.0058. χ_1 = 0.356 and χ_2 = 0.840. Curve 1 ($i = 0$) shows two maxima a and c and one minimum b. Line 2 represents the last term of Equation (11) with $i = 0.22$. Two black dots in Curve 3 ($i = 0.22$) represent the volume fractions of the condensed (ϕ_C) and the swollen (ϕ_S) states in equilibrium with each other.

A situation corresponding to case (c) above may be indicated in Figure 2. Curve 1 in the figure represents a possible swelling curve of the uncharged gel, calculated with Equation (11) with $i = 0$, which shows a transition-like behavior between the swollen state ($\phi = \phi_S$) and the condensed state

($\phi = \phi_C$). When the contribution from $\Delta\mu_w^{ion}/RT$ represented with a straight line 2 is added, curve 3 results, which indicates no transition-like behavior anymore. When the contribution $\Delta\mu_w^{ion}/RT$ is small enough for the second maximum of curve 1 (denoted as c) in Figure 2 to remain positive, the transition-like behavior is preserved even in the presence of electric charges. In the case of NIPA-*co*-AA gels, a threshold of counterion osmotic pressure Π/RT (about 20 mM) was confirmed [9] beyond which the volume change is continuous. As far as the osmotic pressures are below the threshold, the transition-like behavior is preserved. This will be case (b) above. It is to be stated that ϕ-dependent χ parameters, Equation (7), are used in the calculations and prescribed values of χ_1 and χ_2 are chosen so that they correspond to a situation near the critical condition.

It is important, however, to examine the validity of the present model on the basis of the parameter values relevant to each reported experimental result.

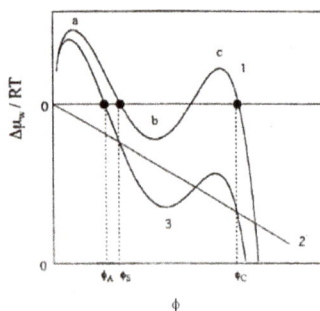

Figure 2. Calculated swelling curves in terms of Equation (11) in the case that the transition behavior disappears when ionic groups are introduced. (v V_1/V_0) = 0.0058. χ_1 = 0.334 and χ_2 = 0.855. Curve 1 (i = 0) represents two maxima a and c and one minimum b. Two black dots represent the volume fractions of the condensed (ϕ_C) and the swollen (ϕ_S) states in equilibrium with each other. Line 2 represents the last term of Equation (11) with i = 0.005. Curve 3 (i = 0.22): a black dot represents the volume fractions of the equilibrium swollen volume fraction (ϕ_A).

A few words are in order with respect to apparently conflicting results on NIPA-*co*-AA gels [6,9,10]. An important factor affecting the swelling behavior of the gel is the pH of the media in which it is immersed. With a decrease in pH, not only the charge density of polymer chains decreases, but also the interaction parameter χ changes due probably to the hydrogen bond formation between COOH and COO$^-$ groups. Different hydrations of the two groups also affect the χ parameter. Accordingly, the charge effect should be examined under the condition where carboxyl groups are nearly completely ionized and the amount of charges should be regulated not by pH, but by changing the amount of acrylic acid (AA) to be introduced. It is to be noted that a similar result as that reported in [6] was obtained in [10] at pH 5.6.

4. Conclusions

Disappearance of the transition-like gel volume change as a result of the introduction of electric charges could be generally understood in terms of the mechanism presented here. In the mechanism it is assumed that the mixing entropy of counterions plays a central role compared with other effects induced by ionic groups such as on the elastic properties and the χ parameter.

Author Contributions: The idea underlying the model presented here was developed through the discussion among Hiroshi Maeda, Shigeo Sasaki, Hideya Kawasaki and Rie Kakehashi. Hiroshi Maeda and Rie Kakehashi wrote the paper.

Conflicts of Interest: The authors declare no conflict of interest.

References

1. Dušek, K.; Patterson, D. Transition in Swollen Polymer Networks Induced by Intramolecular Condensation. *J. Polym. Sci. A-2* **1968**, *6*, 1209–1216. [CrossRef]
2. Tanaka, T. Collapse of Gels and the Critical Endpoint. *Phys. Rev. Lett.* **1978**, *40*, 820–823. [CrossRef]
3. Ilavsky, M.; Hrouz, J.; Ulbrich, K. Phase transition in swollen gels: 3. The Temperature Collapse and Mechanical Behaviour of poly(*N,N*-diethylacrylamide) Networks in Water. *Polym. Bull.* **1982**, *7*, 107–113. [CrossRef]
4. Ilavsky, M.; Hrouz, J.; Havlicek, I. Phase transition in swollen gels: 7. Effect of charge concentration on the temperature collapse of poly(*N,N*-diethylacrylamide) networks in water. *Polymer* **1982**, *7*, 1514–1518. [CrossRef]
5. Hirokawa, Y.; Tanaka, T. Volume Phase Transition in a Non-ionic Gel. *J. Chem. Phys.* **1984**, *81*, 6379–6380. [CrossRef]
6. Hirotsu, S.; Hirokawa, Y.; Tanaka, T. Volume-phase transitions of ionized *N*-isopropylacrylamide gels. *J. Chem. Phys.* **1987**, *87*, 1392–1395. [CrossRef]
7. Beltran, S.; Hooper, H.H.; Blanch, H.W.; Prausnitz, J.M. Swelling equilibria for ionized temperature-sensitive gels in water and in aqueous salt solutions. *J. Chem. Phys.* **1990**, *92*, 2061–2066. [CrossRef]
8. Karbarz, M.; Pulka, K.; Misicka, A.; Stojek, Z. pH and Temperature-Sensitive *N*-isopropylacrylamide Ampholytic Networks Incorporating l-Lysine. *Langmuir* **2006**, *22*, 7843–7847. [CrossRef] [PubMed]
9. Kawasaki, H.; Sasaki, S.; Maeda, H. Effect of Introduced Electric Charge on the Volume Phase Transition of *N*-isopropylacrylamide Gels. *J. Phys. Chem. B* **1997**, *101*, 4184–4187. [CrossRef]
10. Kawasaki, H.; Sasaki, S.; Maeda, H. Effect of pH on the Volume Phase Transition of Copolymer Gels of *N*-isopropylacrylamide and Sodium Acrylate. *J. Phys. Chem. B* **1997**, *101*, 5089–5092. [CrossRef]
11. Erman, B.; Flory, P.J. Critical Phenomena and Transitions in Swollen Polymer Networks and in Linear Macromolecules. *Macromolecules* **1986**, *19*, 2342–2353. [CrossRef]
12. Alexandrowicz, Z. Calculation of the Thermodynamic Properties of Polyelectrolytes in the Presence of Salt. *J. Polym. Sci.* **1962**, *56*, 97–114. [CrossRef]

gels

MDPI

Article

Temperature Dependence of Electrophoretic Mobility and Hydrodynamic Radius of Microgels of Poly(N-isopropylacrylamide)

Yasuyuki Maki [1],* Kentaro Sugawara [2] and Daisuke Nagai [2]

[1] Department of Chemistry, Graduate School of Science, Kyushu University, Fukuoka 819-0395, Japan
[2] Division of Molecular Science, Faculty of Science and Technology, Gunma University, Kiryu 376-8515, Japan; t14801041@gunma-u.ac.jp (K.S.); daisukenagai@gunma-u.ac.jp (D.N.)
* Correspondence: maki@chem.kyushu-univ.jp; Tel.: +81-92-802-4120

Received: 7 March 2018; Accepted: 19 April 2018; Published: 20 April 2018

Abstract: Electrostatic interactions in charged microgels, which are dominated by the microgel net charge, play a crucial role in colloidal stabilization and loading of small, charged molecules. In this study, the temperature dependences of electrophoretic mobility μ and hydrodynamic radius R_h were measured for a slightly ionized poly(N-isopropylacrylamide) (PNIPA) microgel in a dilute suspension. A decrease in R_h was observed in the temperature range between 30 °C and 35 °C, corresponding to the lower critical solution temperature of PNIPA, and an increase in $|\mu|$ was observed in a higher temperature range between 34 °C and 37 °C. The analysis based on electrophoresis theory for spherical polyelectrolytes indicated that the net charge of the microgel decreased as the microgel was deswollen.

Keywords: microgel; electrophoresis; light scattering

1. Introduction

Microgels are colloidal particles of crosslinked polymers swollen by a large amount of water, the size of which ranges from tens of nanometers to several microns [1]. They can reversibly change their swelling degree and, hence, particle size in response to environmental stimuli such as pH value, temperature, and ionic strength [2,3]. This stimulus sensitivity makes microgels excellent candidates for soft materials in biomedical applications such as drug delivery and biosensing [2–4].

Poly(N-isopropylacrylamide) (PNIPA) microgel is one of the most extensively investigated thermosensitive microgels [5]. Aqueous solutions of PNIPA show a lower critical solution temperature (LCST) of 32 °C. The PNIPA microgels deswell above the LCST because of the dehydration of the polymers; the radii of the PNIPA microgels decrease with temperature above the LCST. The simplest method for preparing PNIPA microgel particles is by free radical precipitation polymerization. In this method, N-isopropylacrylamide (NIPA) monomer and a crosslinker are dissolved in water, and the solution is heated above the LCST; then, an initiator is added [6]. The obtained PNIPA microgel shows a narrow size distribution because a precursor particle of the microgel is formed by homogeneous nucleation [7]. For cases in which ionic initiators, such as ammonium persulfate (APS) or potassium persulfate (KPS), are used, the microgels become slightly charged because of the ionized groups imparted by the initiator [7]. Electrostatic interactions in the charged microgel play a crucial role in the colloidal stabilization of the microgel suspension and may also modulate the loading of small, charged drugs by the microgel in the drug delivery system. Thus, it is important to clarify the effect of temperature on the electrostatic properties of the ionized thermosensitive microgels.

Electrophoresis is an electrokinetic phenomenon which reflects both electrostatic and hydrodynamic properties of systems. The electrophoretic behaviors of ionized thermosensitive microgels have been investigated [5,7–12]. These previous studies showed that the absolute value of electrophoretic mobility μ increased with the decrease in the particle radius. This result was partly explained by the increasing charge density, although Daly et al. reported that the onset temperature of the increase in $|\mu|$ was slightly higher than that of the decrease in the particle radius [8]. In previous studies, experimental data of μ as a function of temperature were compared with different theoretical models for a hard sphere with a charged surface [7], a spherical permeable polyelectrolyte [7], and a particle covered with a polyelectrolyte layer [8–11]. These models, however, failed to sufficiently represent the experimental data [5].

In previous analysis of the electrophoretic data [5,7–12], the charge of the microgel has been commonly assumed to be independent of temperature and the particle radius. The charge defined from the number of the ionic groups on the microgel network is referred to as the bare charge, which is intrinsic to the network structure. In contrast, the charge taking into account partial screening inside the particle and/or ion binding of counterions and co-ions is called net charge, or effective charge, which varies with the environment of the microgel. The electrostatic and electrokinetic properties of the microgel are described by the net charge rather than by the bare charge. Recent theoretical and experimental studies have shown that the net charge of the microgel was affected by the swelling degree of the microgel [13–17]. Therefore, it is plausible to reconsider the assumption of the constant charge in the analysis of the electrophoretic data of the thermosensitive microgels as a function of temperature.

In this study, μ and hydrodynamic radius R_h of slightly ionized PNIPA microgels in a dilute suspension were measured as a function of temperature by electrophoretic light scattering (ELS) and dynamic light scattering (DLS), respectively, and the temperature dependence of the net charge of the microgel was investigated. The electrophoretic and hydrodynamic data were interpreted, taking the temperature dependence of the microgel net charge into consideration.

2. Results and Discussion

The characterization of the synthesized PNIPA microgel was carried out by static light scattering (SLS) and DLS measurements at 25.0 °C. The scattering curve of SLS was represented by the scattering equation for a particle with a fuzzy particle surface (Equations (8)–(10) in Materials and Methods) (Figure 1). The solid curve in Figure 1 is the fitted curve according to the equations with parameters shown in Table 1. Relatively small values of the width σ of the smeared particle surface and the polydispersity σ_R of the particle size compared with the average particle radius $<R>$ indicated that the obtained microgel was almost homogeneous and monodisperse. The distribution of R_h obtained from the DLS data (Figure 2) also represented the monodispersity of the sample. The average radius $<R>$ from SLS and the average hydrodynamic radius $<R_h>$ from DLS were 2.8×10^2 nm and 3.0×10^2 nm, respectively, indicating good consistency with each other.

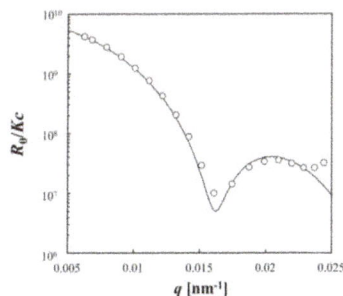

Figure 1. Static light scattering (SLS) data for a dilute suspension of the poly(N-isopropylacrylamide) (PNIPA) microgel measured at 25.0 °C and the fitted curve of the form factor for a particle with a fuzzy particle surface as described in Equations (8)–(10).

Table 1. Static light scattering (SLS) and dynamic light scattering (DLS) results for the PNIPA microgel measured at 25.0 °C.

M_w (10^9 g/mol)	$\langle R \rangle$ (10^2 nm)	σ_R (10^2 nm)	σ (10^2 nm)	$\langle R_h \rangle$ (10^2 nm)
8.2	2.8	0.12	0.29	3.0

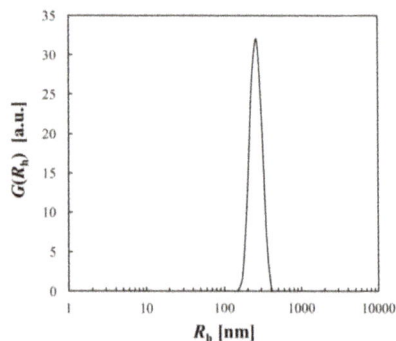

Figure 2. Hydrodynamic radius distribution $G(R_h)$ of the PNIPA microgel measured at 25.0 °C.

The thermosensitive behaviors of the PNIPA microgel were demonstrated in the DLS and ELS experiments (Figure 3). The DLS data showed that as the temperature was increased, $\langle R_h \rangle$ decreased gradually at temperatures below 30 °C, and then a significant decrease in $\langle R_h \rangle$ was observed in the temperature range from 30 °C to 35 °C; subsequently, $\langle R_h \rangle$ decreased slightly at temperatures above 35 °C. This indicates that the noticeable deswelling of the microgel occurred in the temperature range between 30 °C and 35 °C, owing to the dehydration of PNIPA, and the midpoint of the volume change coincided with the LCST of PNIPA (32 °C). In the ELS experiment, the temperature dependence of μ was measured. The negative values of μ indicated that the microgel was negatively charged because of the sulfate group imparted by the initiator APS. The ELS data showed that as the temperature was increased, the absolute value $|\mu|$ of the electrophoretic mobility increased slightly at temperatures below 34 °C, and then a significant increase in $|\mu|$ was observed in the temperature range from 34 °C to 37 °C; subsequently, $|\mu|$ increased gradually at temperatures above 37 °C. The onset temperature of the increase in $|\mu|$ was 4 K higher than that of the decrease in $\langle R_h \rangle$, which supports the findings of a previous study [8].

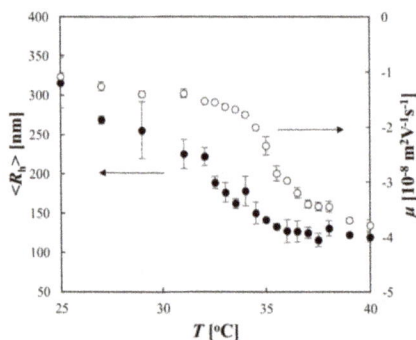

Figure 3. Electrophoretic mobility μ (open circles) and hydrodynamic radius $\langle R_h \rangle$ (filled circles) as a function of temperature for the PNIPA microgel.

In previous studies, experimental data of μ for ionic microgels were discussed in terms of equations based on different models [5,18,19]. One of these equations is based on a simple model of a hard sphere with a charged surface, which is equivalent to the classical Smoluchowski equation [20]. The applicability of this equation is questionable since it is not plausible to assume that all the charges are located on the surface of the microgel, particularly in the highly swollen state at low temperatures [5]. In contrast, the equation derived by Hermans and Fujita is based on a spherical polyelectrolyte model [21]. In this model, charges fixed on the particle are assumed to be uniformly distributed throughout the particle volume, and the particle is porous and permeable to small ions. The polyelectrolyte model has been used for the analysis of electrophoretic data in many experimental studies [7,18,19]. More recently, a set of equations describing μ for solid colloidal particles covered with an ion-penetrable layer of polyelectrolytes was derived by Ohshima [22,23]. This model has been used for analyzing μ of microgels in recent experimental studies [8–11] because it was proposed that microgels prepared by precipitation polymerization contained a lightly crosslinked particle periphery [24], and more charges were located in the peripheral region of the microgel in analogy to persulfate-initiated emulsifier-free polystyrene latex [7]. However, the thickness of the polyelectrolyte layer in this model cannot be determined directly from the experimental data and this value has been estimated based on various assumptions in previous studies [8,10,12]. In the present study, μ of the PNIPA microgel was discussed in terms of the spherical polyelectrolyte model in which no assumptions on the thickness of the polyelectrolyte layer are needed, although it is difficult to conclude which model is more suitable in the present case. In contrast with polystyrene latex, the core of PNIPA microgels should be hydrophilic even at the temperature used during the synthesis, and many sulfate groups could be located in the particle interior [7]. Moreover, even though the PNIPA microgel was prepared by precipitation polymerization in the present study, SLS data showed that the obtained microgel was a relatively homogeneous particle with a thin smeared surface. Thus, we suggest that the polyelectrolyte model is adequate for analyzing μ of the microgel in the present case.

In the previous studies, the temperature dependence of μ was not well represented by the theories described above when the charge of the microgel was assumed to be constant [5,7–12]. In the present study, the net charge of the microgel was calculated as a function of temperature by comparing the data of μ with the Hermans–Fujita equation for the polyelectrolyte model. According to the theory of Hermans and Fujita [21], the μ of spherical polyelectrolytes is described as

$$\mu = \frac{zev_c}{\eta\alpha^2}\left[1 + \frac{2}{3}\left(\frac{\alpha}{\kappa}\right)^2 \frac{1 + \alpha/(2\kappa)}{1 + \alpha/\kappa}\right] \tag{1}$$

where z, e, and v_c represent the valence of the ionic group on the microgel network, the elementary electric charge, and the effective number of monomer units with charged moiety from the initiator per unit volume, respectively. The net charge Z is related to v_c by $Z = (4\pi/3) v_c <R_h>^3$. In Equation (1), κ is the Debye–Hückel parameter defined as

$$\kappa = \sum_i \left(\frac{\rho_i^\infty e^2 z_i^2}{\varepsilon_0 \varepsilon k T}\right)^{1/2} \tag{2}$$

where ρ_i^∞ and z_i are the bulk concentration and the valence of the ith species of mobile ions, respectively; ε_0, ε, k, and T are the permittivity of a vacuum, the relative permittivity, the Boltzmann constant, and the absolute temperature, respectively. The parameter α in Equation (1) is defined as

$$\alpha = \sqrt{\gamma/\eta} \tag{3}$$

where η is the viscosity and γ represents a frictional coefficient of the microgel per unit volume. Given that the microgel can be regarded as a porous medium and the frictional force is exerted by spheres of radius b corresponding to monomer units which are uniformly distributed in the microgel, γ is given by [25,26]

$$\gamma = 6\pi\eta b\nu F(\varphi) \tag{4}$$

where ν is the number density of monomers and $F(\varphi)$ is a drag coefficient as a function of the volume fraction of polymer [10,26]. Koch et al. showed that $F(\varphi)$ for $\varphi < 0.4$ and $\varphi \geq 0.4$ can be represented by

$$F(\varphi) = \frac{1 + 3(\varphi/2)^2 + (135/64)\varphi \ln \varphi + 16.456\varphi}{1 + 0.681\varphi - 8.48\varphi^2 + 8.16\varphi^3} \qquad (\varphi < 0.4) \tag{5}$$

and

$$F(\varphi) = \frac{10\varphi}{(1-\varphi)^3} \qquad (\varphi \geq 0.4) \tag{6}$$

respectively [27]. From Equation (5), $F(\phi) = 1$ as $\phi \to 0$. Using the relation $\phi = (4\pi/3) \, b^3 \, \nu$ and Equations (5) and (6), ϕ and $F(\phi)$ were calculated from the data of $<R_h>$ in Figure 3 and are shown as a function of temperature in Figure 4. Here, the monomer density ν was calculated by the relation $\nu = (M_w/M_0)/(4\pi<R_h>^3/3)$, where M_0 is the molecular weight of the NIPA monomer (113.16). The value of b was estimated to be 3.4 Å from the equation $b = (3v_0M_0/4\pi N_A)^{1/3}$, where N_A is Avogadro's number; as the specific volume v_0 of the monomer, 0.862 cm^3/g was used for NIPA.

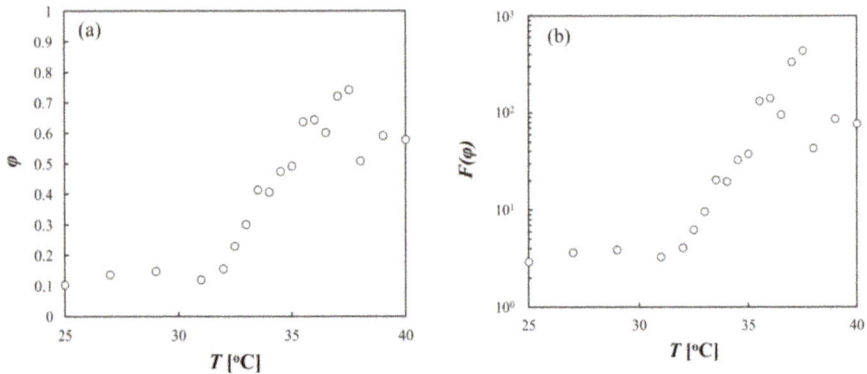

Figure 4. (a) Volume fraction ϕ of polymer as a function of temperature; (b) drag coefficient $F(\phi)$ as a function of temperature.

From the data of μ and $<R_h>$ in Figure 3, the fraction of effectively charged monomers ν_c/ν was obtained by using Equations (1)–(6) (Figure 5). The value of (ν_c/ν) reflects the net charge Z of the microgel because of the relation $Z = (\nu_c/\nu) (M_w/M_0)$. Figure 5 shows that as the temperature increased, ν_c/ν was almost constant at temperatures below 31 °C, and it decreased significantly near the LCST (in the temperature range from 31 °C to 33 °C); it was then nearly constant at temperatures above 33 °C. This indicates that the net charge of the microgel decreased when the microgel deswelled. We suggest that this change in the net charge was mainly due to the change in the electrostatic property of the microgel, and the estimation of the net charge was not affected by the change in the hydrodynamic property; when $F(\phi) = 1$ was used for all temperatures, the obtained values of ν_c/ν were almost the same as those shown in Figure 5 (data not shown).

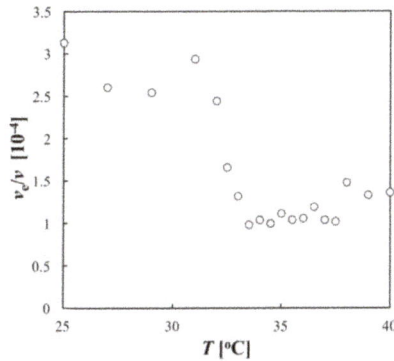

Figure 5. Fraction of effectively charged monomers v_c/v as a function of temperature.

Figure 3 showed that the onset temperature of the increase in $|\mu|$ was higher than that of the decrease in $<R_h>$. Daly et al. previously observed the same phenomenon and, on the basis of the analysis of μ with Ohshima's equation, they attributed this deviation in the onsets to a three-stage deswelling process: core collapse in the first stage, partial shell collapse in the second stage, and further core collapse in the third stage [8]. However, the mechanism for this complicated deswelling process was not clarified in their study. Considering the temperature dependence of the net charge of the microgel shown in the present study, this phenomenon was explained simply as follows. In the present experimental condition, $\kappa^{-1} \sim 3 \times 10^{-8}$ m and α varied between 3×10^9 m and 4×10^{10}. Therefore, α/κ was sufficiently large for the first term in the bracket in Equation (1) to be regarded as negligible relative to the second term and for an approximation of $(1 + \alpha/2\kappa)/(1 + \alpha/\kappa) \sim 1/2$ to be used. This gives the following equation:

$$\mu = \frac{ze}{4\pi\eta b^3 \kappa^2} \left(\frac{v_c}{v}\right)\varphi. \tag{7}$$

Here, the relation $\phi = (4\pi/3)\, b^3\, v$ was used. Equation (7) shows that $|\mu|$ is increased by increasing ϕ, but is decreased by decreasing v_c/v. The gradual change in $|\mu|$ with temperature in the range of 31 °C to 33 °C occurs because ϕ increased approximately threefold (Figure 4) and v_c/v decreased approximately threefold (Figure 5). In contrast, $|\mu|$ increased significantly with temperature in the range of 33 °C to 37 °C because ϕ increased further but v_c/v was almost constant. Thus, the increase in the charge density of the microgel due to increasing ϕ was cancelled out by the decrease in v_c/v, or the net charge Z in the specific temperature range, resulting in the deviation of the onset temperature of $|\mu|$ from that of $<R_h>$.

Figure 5 shows that the net charge of the microgel decreases with the decrease in the swelling degree of the microgel. The effect of the swelling degree of the microgel on its net charge is not fully understood. Theoretical studies in which the excluded volume repulsion between the microgel and the ions was explicitly taken into account showed that the net charge of the microgel increased with decreasing particle radius because the ions were expelled from the inside of the microgel in the deswollen state, which led to less efficient screening of the microgel charge [13,14]. In contrast, a more mathematically rigorous approach showed that the net charge was decreased by the deswelling of the microgel because the short-ranged association between the charged microgel and the neighboring small ions became stronger, mainly because of the increasing bare charge density of the microgel [15]. The present study and other experiments of light scattering, phase behavior, and conductivity for microgel suspensions [16,17] support the latter theoretical prediction. For PNIPA microgel in the deswollen state, the environment inside the particle should be more hydrophobic because of the dehydration of the polymers, which may affect effective permittivity around the ionic groups in the microgel. It has been reported for various biopolymers that hydrophobic groups in the vicinity of an ionic group prevent dissociation of the ionic group [28,29]. The change in the dissociation constant of the ionic group could be also attributed to the observed decrease in the net charge of the microgel at high temperatures.

3. Conclusions

The μ and $<R_h>$ of slightly charged PNIPA microgel were measured as functions of temperature. The temperature dependence of μ was interpreted in terms of the change in the net charge of the microgel by analysis based on the spherical polyelectrolyte model. It was shown that the net charge of the microgel decreased significantly as $<R_h>$ sharply decreased near the LCST of PNIPA. This finding is important for designing colloidal stability of microgel suspensions and drug loading efficiency in microgel-based drug delivery systems.

4. Materials and Methods

4.1. Preparation of Microgels

The NIPA, N,N'-methylenebisacrylamide (MBA) as a crosslinker, and APS as an initiator were obtained from Wako Pure Chemical Industries, Ltd., Osaka, Japan. The NIPA was recrystallized in a toluene/hexane mixture.

In a two-necked round-bottom flask, equipped with a stirrer and a reflux condenser, NIPA (0.564 g, 4.98 mmol) and MBA (0.0213 g, 0.138 mmol) were dissolved in 40 mL distilled water under stirring. After bubbling the solution with nitrogen, the solution was heated to 70 °C under a nitrogen purge and APS (0.0271 g, 0.119 mmol) dissolved in 1.2 mL water was then added to the monomer solution. The polymerization was continued for 24 h and then cooled down to room temperature. The obtained dispersion of microgels was purified by dialysis in Milli-Q water for two weeks and diluted for further measurements as described in the following.

4.2. SLS, DLS, and ELS Measurements

For the SLS experiment, a dilute dispersion of the microgels in 0.1 mM NaCl at a polymer concentration of $c = 0.98 \times 10^{-5}$ g/cm^3 was prepared. A laboratory-made instrument equipped with a He–Ne laser (632.8 nm wavelength) and a photomultiplier on a goniometer were used for the SLS measurements. In the measurements, the scattered intensity in the angular range from 27.5° to 135° was obtained as a function of the scattering vector length $q = (4\pi n/\lambda) \sin(\theta/2)$, where n, λ, and θ are the refractive index of the solution, the wavelength of the incident light in a vacuum, and the scattering angle, respectively. The SLS data were analyzed on the basis of the form factor $P(q)$ for an inhomogeneous particle with a fuzzy particle surface [30,31] as

$$P(q) = \left(\frac{3[\sin(qR) - qR\cos(qR)]}{(qR)^3} e^{-(q\sigma)^2/2} \right)^2 \tag{8}$$

where R and σ are the radius of the particle and the width of the smeared particle surface, respectively. Taking into account the polydispersity of the particles, the excess Rayleigh ratio R_θ measured at a scattering angle θ as a function of q was fitted by the following equation:

$$\frac{R_\theta}{Kc} = M_w \int_0^\infty P(q, R) D(R, R, \sigma_R) dR \tag{9}$$

where K is defined as $K = (4\pi^2 n^2/N_A \lambda^4)(dn/dc)^2$, in which (dn/dc) is the refractive index increment and was determined to be 0.182 cm^3/g by measurements with a differential refractometer; M_w is the weight-averaged molecular weight; and $D(R, <R>, \sigma_R)$ is a Gaussian distribution function with respect to the particle radius R given by

$$D(R, \langle R \rangle, \sigma_R) = \frac{1}{\sqrt{2\pi}\sigma_R} \exp\left[-\frac{(R - R)^2}{2\sigma_R^2}\right] \tag{10}$$

where <R> is the average particle radius and σ_R represents the polydispersity of the particle size.

For the DLS and ELS experiments, a dilute dispersion of the microgels in 0.1 mM NaCl at $c = 2.0 \times 10^{-5}$ g/cm^3 was prepared. The DLS and ELS measurements were performed with a Malvern Zetasizer Nano ZS instrument with a He–Ne laser at a wavelength of 632.8 nm at a fixed scattering angle of 13°. In the DLS measurement, the z-average diffusion coefficient D_z was obtained by the cumulant analysis of the autocorrelation function, and the <R_h> was calculated using the Stokes–Einstein equation:

$$\langle R_h \rangle = \frac{kT}{6\pi\eta D_z} . \tag{11}$$

The hydrodynamic radius distribution $G(R_h)$ was also obtained from the autocorrelation function by CONTIN analysis. In the ELS measurements, μ was obtained by the phase analysis light scattering (PALS) method [32].

Acknowledgments: This work was partly supported by JSPS KAKENHI (Grants-in-Aid for Scientific Research) Grant Number 15K20906.

Author Contributions: Yasuyuki Maki and Daisuke Nagai conceived and designed the experiments; Kentaro Sugawara performed the experiments and analyzed the data; Yasuyuki Maki wrote the paper.

Conflicts of Interest: The authors declare no conflict of interest.

References

1. Saunders, B.R.; Vincent, B. Microgel particles as model colloids: Theory, properties and applications. *Adv. Colloid Interface Sci.* **1999**, *80*, 1–25. [CrossRef]
2. Thorne, J.B.; Vine, G.J.; Snowden, M.J. Microgel application and commercial consideration. *Colloid Polym. Sci.* **2011**, *289*, 625–646. [CrossRef]
3. Saunders, B.R.; Laajam, N.; Daly, E.; Teow, S.; Hu, X.; Stepto, R. Microgels: From responsive polymer colloids to biomaterials. *Adv. Colloid Interface Sci.* **2009**, *147–148*, 251–262. [CrossRef] [PubMed]
4. Guan, Y.; Zhang, Y. PNIPAM microgels for biomedical applications: From dispersed particles to 3D assemblies. *Soft Matter* **2011**, *7*, 6375–6384. [CrossRef]
5. Pelton, R. Temperature-sensitive aqueous microgels. *Adv. Colloid Interface Sci.* **2000**, *85*, 1–33. [CrossRef]
6. Pelton, R.H.; Chibante, P. Preparation of aqueous latices with N-isopropylacrylamide. *Colloid Surf.* **1986**, *20*, 247–256. [CrossRef]
7. Pelton, R.H.; Pelton, H.M.; Morphesis, A.; Rowell, R.L. Particle sizes and electrophoretic mobilities of poly(N-isopropylacrylamide) latex. *Langmuir* **1989**, *5*, 816–818. [CrossRef]
8. Daly, M.; Saunders, B.R. Temperature-dependent electrophoretic mobility and hydrodynamic radius measurements of poly(N-isopropylacrylamide) microgel particles: Structural insights. *Phys. Chem. Chem. Phys.* **2000**, *2*, 3187–3193. [CrossRef]
9. Ramusson, M.; Vincent, B.; Marston, N. The electrophoresis of poly(N-isopropylacrylamide) microgel particles. *Colloid Polym. Sci.* **2000**, *278*, 253–258. [CrossRef]
10. García-Salinas, M.J.; Romero-Cano, M.S.; de las Nieves, F.J. Electrokinetic characterization of poly(N-isopropylacrylamide) microgel particles: Effect of electrolyte concentration and temperature. *J. Colloid Interface Sci.* **2001**, *241*, 280–285. [CrossRef] [PubMed]
11. López-León, T.; Ortega-Vinuesa, J.L.; Bastos-González, D.; Elaïssari, A. Cationic and anionic poly(N-isopropylacrylamide) based submicron gel particles: Electrokinetic properties and colloidal stability. *J. Phys. Chem. B* **2006**, *110*, 4629–4636. [CrossRef] [PubMed]
12. Sierra-Martín, B.; Romero-Cano, M.S.; Fernández-Nieves, A.; Fernández-Barbero, A. Thermal control over the electrophoresis of soft colloidal particles. *Langmuir* **2006**, *22*, 3586–3590. [CrossRef] [PubMed]
13. Moncho-Jordá, A. Effective charge of ionic microgel particles in the swollen and collapsed states: The role of the steric microgel-ion repulsion. *J. Chem. Phys.* **2013**, *139*, 064906. [CrossRef] [PubMed]

14. Adroher-Benítez, I.; Ahualli, S.; Bastos-González, D.; Ramos, J.; Forcada, J.; Moncho-Jordá, A. The effect of electrosteric interactions on the effective chare of thermresponsive ionic microgels: Theory and experiments. *J. Polym. Sci. Part B Polym. Phys.* **2016**, *54*, 2038–2049. [CrossRef]
15. González-Mozuelos, P. Effective electrostatic interactions among charged thermo-responsive microgels immersed in a simple electrolyte. *J. Chem. Phys.* **2016**, *144*, 054902. [CrossRef] [PubMed]
16. Braibanti, M.; Haro-Pérez, C.; Quesada-Pérez, M.; Rojas-Ochoa, L.F.; Trappe, V. Impact of volume transition on the net charge of poly-*N*-isopropyl acrylamide microgels. *Phys. Rev. E* **2016**, *94*, 032601. [CrossRef] [PubMed]
17. Holmqvist, P.; Mohanty, P.S.; Nägele, G.; Schurtenberger, P.; Heinen, M. Structure and dynamics of loosely cross-linked ionic microgel dispersions in the fluid regime. *Phys. Rev. Lett.* **2012**, *109*, 048302. [CrossRef] [PubMed]
18. Ogawa, K.; Nakayama, A.; Kokufuta, E. Electrophoretic behavior of ampholytic polymers and nanogels. *J. Phys. Chem. B* **2003**, *107*, 8223–8227. [CrossRef]
19. Kokufuta, E.; Sato, S.; Kokufuta, M.K. Electrophoretic behavior of microgel-immobilized polyions. *Langmuir* **2013**, *29*, 15442–15449. [CrossRef] [PubMed]
20. Hunter, R.J. *Zeta Potential in Colloid Science*; Academic Press: London, UK, 1981; pp. 69–75.
21. Hermans, J.J.; Fujita, H. Electrophoresis of charged polymer molecules with partial free drainage. *Proc. Akad. Amst.* **1955**, *58*, 182–187.
22. Ohshima, H. Electrophoretic mobility of soft particles. *J. Colloid Interface Sci.* **1994**, *163*, 474–483. [CrossRef]
23. Ohshima, H. On the general expression for the electrophoretic mobility of a soft particle. *J. Colloid Interface Sci.* **2000**, *228*, 190–193. [CrossRef] [PubMed]
24. Wu, X.; Pelton, R.H.; Hamielec, A.E.; Woods, D.R.; McPee, W. The kinetics of poly(*N*-isopropylacrylamide) microgel latex formation. *Colloid Polym. Sci.* **1994**, *272*, 467–477. [CrossRef]
25. Ohshima, H. Electrokinetics of soft particles. *Colloid Polym. Sci.* **2007**, *285*, 1411–1421. [CrossRef]
26. Ohshima, H. Electrophoresis and electrostatic interaction of soft particles. *Butsuri* **2013**, *68*, 89–97.
27. Koch, D.L.; Sangani, A.S. Particle pressure and marginal stability limits for a homogeneous monodispersed gas-fluidized bed: Kinetic theory and numerical simulations. *J. Fluid Mech.* **1999**, *400*, 229–263. [CrossRef]
28. Gekko, K.; Noguchi, H. Potentiometric studies of hydrophobic effect on ion binding of ionic dextran derivatives. *Biopolymers* **1975**, *14*, 2555–2565. [CrossRef]
29. Mehler, E.L.; Fuxreiter, M.; Simon, I.; Garcia-Moreno, E.B. The role of hydrophobic microenvironments in modulating pKa shifts in proteins. *Proteins Struct. Funct. Genet.* **2002**, *48*, 283–292. [CrossRef] [PubMed]
30. Stieger, M.; Richtering, W.; Pederson, J.S.; Lindner, P. Small-angle neutron scattering study of structural changes in temperature sensitive microgel colloids. *J. Chem. Phys.* **2004**, *120*, 6197–6206. [CrossRef] [PubMed]
31. Meyer, S.; Richtering, W. Influence of polymerization conditions on the structure of temperature-sensitive poly(*N*-isopropylacrylamide) microgels. *Macromolecules* **2005**, *38*, 1517–1519. [CrossRef]
32. Miller, J.F.; Schätzel, K.; Vincent, B. The determination of very small electrophoretic mobilities in polar and nonpolar colloidal dispersions using phase analysis light scattering. *J. Colloid Interface Sci.* **1991**, *143*, 532–554. [CrossRef]

gels

MDPI

Article

Irreversible Swelling Behavior and Reversible Hysteresis in Chemically Crosslinked Poly(vinyl alcohol) Gels

Keiichiro Kamemaru [1], Shintaro Usui [2], Yumiko Hirashima [1] and Atsushi Suzuki [2],*

[1] Graduate School of Education, Yokohama National University, 79-2 Tokiwadai, Hodogaya-ku,
    Yokohama 240-8501, Japan; kamemaru@pen-kanagawa.ed.jp (K.K.); hirashima-yumiko-mc@ynu.ac.jp (Y.H.)
[2] Graduate School of Environment and Information Sciences, Yokohama National University, 79-7 Tokiwadai,
    Hodogaya-ku, Yokohama 240-8501, Japan; usui.shintarou@ma.nichigo.co.jp
*   Correspondence: asuzuki@ynu.ac.jp; Tel./Fax: +81-45-339-3846

Received: 14 April 2018; Accepted: 14 May 2018; Published: 21 May 2018

Abstract: We report the swelling properties of chemically crosslinked poly(vinyl alcohol) (PVA) gels with high degrees of polymerization and hydrolysis. Physical crosslinking by microcrystallites was introduced in this chemical PVA gel by a simple dehydration process. The equilibrium swelling ratio was measured in several mixed solvents, which comprised two-components: a good solvent (water or dimethyl sulfoxide (DMSO)), and a poor organic solvent for PVA. In the case of aqueous/organic solvent mixtures subjected to a multiple-sample test, the swelling ratio decreased continuously when the concentration of the organic solvent increased, reaching a collapsed state in the respective pure organic solvents. In the case of DMSO, starting from a swollen state, the swelling ratio rapidly decreased by between 15 and 50 mol % when the concentration of the organic compound increased in a single-sample test. To understand the hysteresis phenomenon, the swelling ratio was measured in a DMSO/acetone mixed solvent, starting from a collapsed state in acetone. The reversibility of swelling in response to successive concentration cycles between DMSO and acetone was examined. As a result, an irreversible swelling behavior was observed in the first cycle, and the swelling ratio in acetone after the first cycle became larger than the initial ratio. Subsequently, the swelling ratio changed reversibly, with a large hysteresis near a specific molar ratio of DMSO/acetone of 60/40. The microstructures were confirmed by Fourier transform infrared spectroscopy during the cycles. The irreversible swelling behavior and hysteresis are discussed in terms of the destruction and re-formation of additional physical crosslinking in the chemical PVA gels.

Keywords: poly(vinyl alcohol); chemical gel; microcrystallite; hydrogen bond; swelling behavior; hysteresis

---

1. Introduction

Hydrogels are defined as three-dimensional, crosslinked polymer networks swollen in water, for which an infinitesimal change in environment can bring on a large volume change [1]. The backbone networks are usually formed through the chemical or physical crosslinking of polymers. In chemical hydrogels, polymer chains are connected by covalent bonds, whereas in physical hydrogels, the bonds are non-covalent. Among synthetic polymers, both chemical and physical hydrogels of poly(vinyl alcohol) (PVA) have been extensively studied for practical applications, owing to their low toxicity and high biocompatibility [2–4].

Physical PVA gels can be prepared easily by a freeze-thaw method [5–7] in water, or by a cast-drying method [8,9], starting from an aqueous PVA solution. Swelling and mechanical properties depend on the microcrystallites, which can be examined using X-ray diffraction (XRD) and Fourier transform infrared (FT-IR) spectroscopy [5,7,10]. On the other hand, chemical PVA gels can be obtained by crosslinking PVA in solution by irradiation with γ-rays or electron beams [11,12], or by chemical reaction using crosslinking agents [13,14].

The swelling behavior of chemical PVA gels slightly crosslinked by glutaraldehyde (GA) [14–16] in mixed solvents was reported. It was found that the microcrystallites could be formed in a chemical PVA gel with high degrees of polymerization and hydrolysis during a dehydration process, which was confirmed by XRD measurements and FT-IR spectroscopy [16,17]. The formation of microcrystallites was suppressed by increasing the degree of chemical crosslinking. The swelling ratio in pure water at room temperature depends strongly on the degree of chemical crosslinking, resulting from the destruction of microcrystallites. In the case of a lower degree of chemical crosslinking density, the gel swells to a size larger than the initial volume at gelation. Moreover, the swelling ratios of the chemical PVA gels with different degrees of polymerization and hydrolysis have been measured in mixed solutions of water and organic solvents. Chemical PVA gels with high degrees of polymerization and hydrolysis assume a swollen state both in pure water and pure dimethyl sulfoxide (DMSO) (good solvents for PVA) [18], whereas these gels collapse in acetone, methanol, or ethanol (poor solvents). In addition to these simple swelling-shrinking changes, the cosolvent and cononsolvent behaviors are observed, depending on the degree of hydrolysis, temperature, and the combination of solvents [15]. In these experiments of solvent dependence, mixed solvents were prepared at different concentrations, and the samples were immersed in the respective solvents using a multiple-sample test. In order to discuss swelling behavior however, solvent concentration should be changed using one sample (single-sample test), which has not been extensively reported so far.

In this study, PVA gels with high degrees of polymerization and hydrolysis were prepared by chemically crosslinking PVA using glutaraldehyde. Physical crosslinking by microcrystallites was introduced into this chemical PVA gel by a simple dehydration process. The swelling behavior of this chemical gel with physical crosslinks was examined in a two-component solvent: one is a good solvent (water or DMSO), while the other is poor (organic compound) for PVA. When the dehydrated sample was immersed in a solvent, it absorbed the solvent, swelled, and remained in its equilibrium gel state, which resulted in the destruction of weak microcrystallites. Firstly, the swelling ratio in aqueous/organic solvent mixtures with different ratios was measured by the multiple-sample test. Secondly, in order to understand the reversibility of the swelling ratio by changing the solvent composition, the swelling behavior was examined in mixed solvents of DMSO and organic solvents by the single-sample test, starting from an equilibrium state in DMSO. Finally, to explore the possibility of realizing a volume phase transition, the swelling ratio was measured in a DMSO/acetone mixed solvent using a single-sample test as a typical example of a combination of good and poor solvents. The swelling behavior is discussed in terms of the network microstructure of the gel, which was examined here by FT-IR measurements.

2. Results and Discussion

2.1. Swelling Behavior in Aqueous/Organic Mixtures

The swelling ratio, d/d_0, of the chemical PVA gel in water was measured in several aqueous/organic solvent mixtures by a multiple-sample test. The gels were dehydrated after gelation and placed in their respective solvents. As shown in Figure 1, at lower concentrations of the organic solvents (alcohols, acetone, dioxane, formamide, ethanol, or DMSO) d/d_0 was dependent on the solvent. At higher concentrations, d/d_0 decreased continuously with an increase in concentration, reaching the collapsed state at 0.4 in the respective pure organic solvent (100 mol %), except in the case of DMSO. In the cases of propanols, d/d_0 first increased and began to decrease from 10 mol %, before finally reaching a collapsed state. This exceptional change can be attributed to structural changes of bound water to the polymer [19]. The swelling behavior can be understood in terms of not only the solubility between the polymers and organic solvents, but also the formation and destruction of physical crosslinkings in the gels.

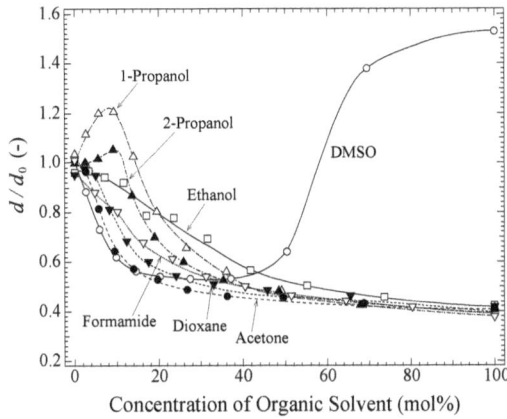

Figure 1. Swelling ratio of chemically crosslinked and dehydrated gels with different ratios of aqueous/organic solvent mixtures at room temperature (25 °C) measured by a multiple-sample test.

It should be noted that the gels showed a collapsed swelling ratio of $d/d_0 = 0.4$ in all the aqueous/organic solvent mixtures studied here, except for DMSO, i.e., for all two-component solvent mixtures containing a good solvent (water) and a poor solvent (organic). Considering that DMSO is a good organic solvent for PVA, and that d/d_0 (=1.6) is much larger than that in pure water ($d/d_0 = 1.0$), it would be interesting to measure d/d_0 in a mixed solvent containing a poor organic solvent and DMSO. Starting from an equilibrium state in DMSO (organic solvent = 0 mol %), in which a dehydrated sample of the chemical PVA gel was placed, d/d_0 was measured by a single-sample test. As is shown in Figure 2, d/d_0 initially decreased gradually while increasing the organic solvent concentration, and it rapidly decreased at between 30 and 50 mol %. In order to examine the detailed behavior, d/d_0 was measured by a single-sample test using DMSO/acetone as a typical example of the two-component system of organic solvent and DMSO; this has been reported in the next section.

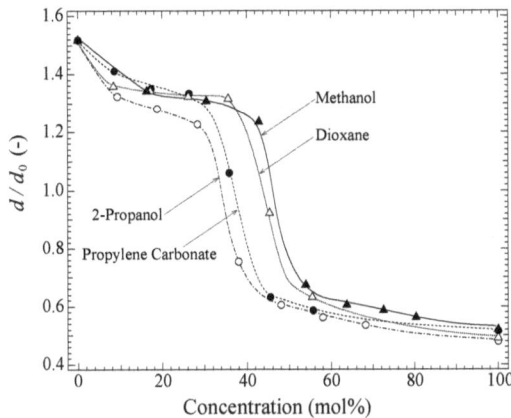

Figure 2. Swelling ratio of chemically crosslinked and dehydrated gels in DMSO-organic solvent mixtures with increasing concentration from single-sample tests.

2.2. Swelling Ratio in DMSO/Acetone Mixed Solvent

The equilibrium swelling ratio, d/d_0 in a DMSO/acetone mixed solvent, was measured by varying successively the acetone concentration, which initially corresponded to the collapsed state of acetone ($d/d_0 = 0.4$). Figure 3 shows the equilibrium d/d_0 as a function of the acetone concentration at 25 °C. During the first decreasing process, where the acetone concentration was decreased, the system remained in the collapsed state above 70%, then it started to increase gradually, before becoming more rapid between 30 and 20 mol %, reaching the swollen state in pure DMSO. During the first increasing process, where the acetone concentration was increased again, d/d_0 initially decreased gradually, and then rapidly between 40 and 50 mol %, and the gel returned to the collapsed state in pure acetone. As seen in this figure, an irreversible swelling behavior was observed, and the swelling ratio in acetone after cycling ($d/d_0 = 0.5$) became larger than the initial ratio. Hereafter, the acetone concentration was again decreased (the second decreasing process) to that of pure DMSO, and increased continuously (the second increasing process), followed by the third decreasing process. The swelling ratio changed continuously and reversibly, with a large hysteresis at near the specific molar ratio of DMSO/acetone = 60/40.

The diameter change of the gels was continuous, and a large hysteresis was observed near a specific molar ratio. During the decreasing and increasing processes, the gels displayed reversible swelling behavior, except for the first decreasing process.

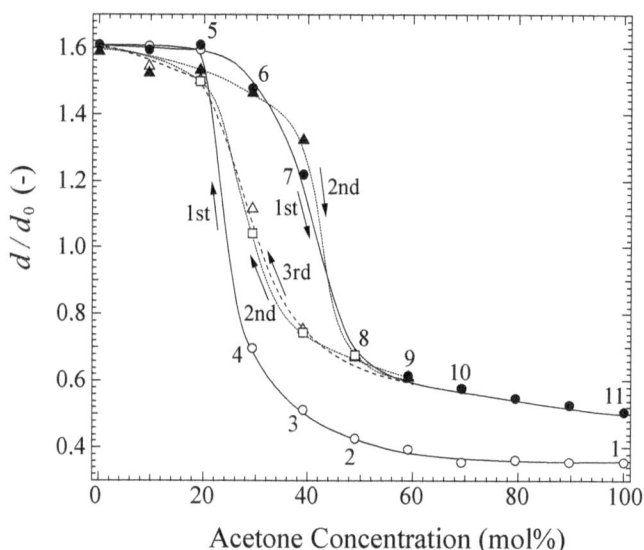

Figure 3. Swelling ratio of chemically crosslinked and dehydrated gels in a DMSO/acetone mixed solvent obtained by a single-sample test. The measurement was initiated from the collapsed state in acetone, and the concentration was successively changed relative to DMSO and acetone.

2.3. Change in Microcrystallites in the DMSO/Acetone Mixed Solvent

In order to discuss the effects of destruction and re-formation of microcrystallites on swelling behavior, ATR FT-IR spectra of the thin plate gel in the respective solvents were obtained during the first decreasing and successive increasing processes. The experimental conditions used for this thin plate gel were the same as those for the thin cylindrical gels used in the above diameter measurements.

Figure 4a shows the IR spectrum of gels with solvent during the first decreasing and successive increasing processes, which correspond to the processes shown in Figure 3. The spectra were normalized based on the C–H bending vibration at 1427 cm^{-1}. Under these present experimental conditions, the peak area at ca. 1427 cm^{-1}, which can be assigned to the –CH$_2$ bending with the deformation bands of C–CH$_3$ appearing at ca. 1377 cm^{-1} [20], was assumed to be equivalent in all spectra, and the spectra were normalized by the respective peak areas. The IR spectra of pure DMSO and acetone are added to this figure. The peaks located at ca. 1360 and 1210 cm^{-1} originate in the presence of acetone, while those at ca. 1050 and 900 cm^{-1} in the presence of DMSO. The large peak at around 3300 cm^{-1} consists of two components, which is assigned to O–H stretching of the non-hydrogen-bonded (3409 cm^{-1}) and hydrogen-bonded O–H groups (3295 cm^{-1}) [17]. As shown in this figure, the peak shifts to higher wavenumbers in the first decreasing process (No. 1 to 5), and to lower wavenumbers in the first increasing process (No. 5 to 11). The peak height shows a tendency to decrease and then increase during this cycle. These observations indicate that the ratio of the numbers of free to hydrogen bonded O–H groups increased and decreased, since the hydrogen bonds between polymer chains were respectively destroyed and re-formed by changing the DMSO/acetone ratio. On the other hand, the peak at 1143 cm^{-1} corresponding to C–O stretching due to the microcrystallites [21–23] disappears in the swollen state and re-appears in the collapsed phase. Figure 4b shows that the peak area at 1143 cm^{-1} was obtained by deconvoluting the spectra to remove the effects of the other peaks. The peak area at 1143 cm^{-1} decreases in the first decreasing process and increases in the successive increasing process, resulting from the decrement and increment of the ratio of microcrystallites to amorphous networks, or in other words, the decrement and increment in the degree of crystallization, respectively. There seems to be a discrepancy between the swelling ratio and peak area. For example, the peak area of No. 4 is almost zero in Figure 4b, although the corresponding swelling ratio does not reach the swollen state. This discrepancy might result from deficiencies in the experimental conditions, such as those due to the evaporation of solvent water at the surface.

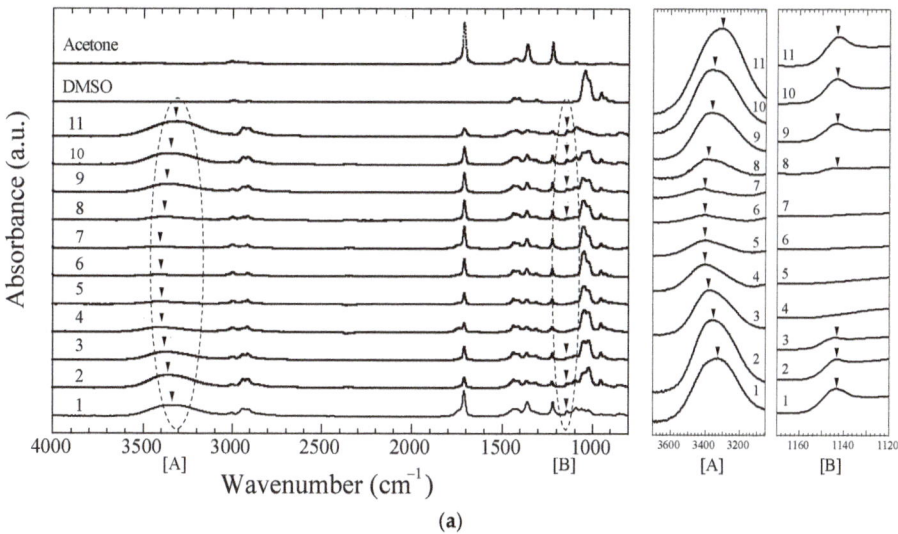

(a)

Figure 4. *Cont.*

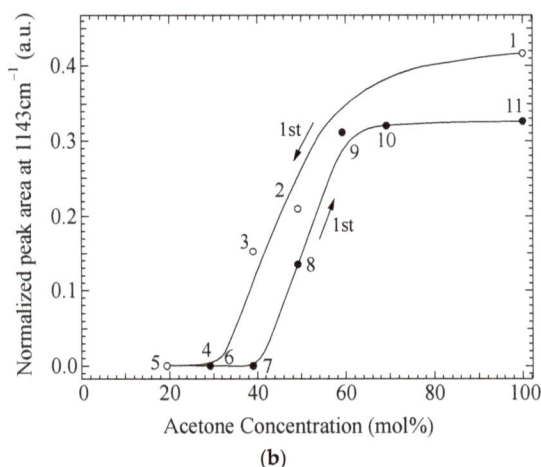

Figure 4. (**a**) ATR FT/IR spectra of a chemically crosslinked and dehydrated gel in various DMSO/acetone mixtures during the single-sample test shown in Figure 3. The detailed changes at around 3300 cm^{-1} (region [A]) and 1143 cm^{-1} (region [B]) are displayed; (**b**) The peak area at 1143 cm^{-1} normalized by the peak area at ca. 1427 cm^{-1}, obtained by deconvoluting the spectra to remove the effects of other peaks. The respective number from 1 to 11 corresponds to those in Figure 3. The acetone concentrations are as follows; No. 1: 100, No. 2: 50, No. 3: 40, No. 4: 30, No. 5: 20, No. 6: 30, No. 7: 40, No. 8: 50, No. 9: 60, No. 10: 70, No. 11: 100 mol %.

2.4. Origin of Reversible Change with a Large Hysteresis

Two important observations can be made about the observed swelling behavior of chemically crosslinked PVA gels in response to solvent concentration changes in DMSO/acetone mixed solutions. One is that there is an irreversible swelling behavior during the first decreasing and increasing processes. This irreversible swelling behavior suggests that microcrystallites are destroyed during the first decreasing process, and they were not completely re-formed during the first increasing process. In other words, microcrystallites can recover, but the recovered amount is smaller than the initial amount. According to a report on the ionized gels of poly(*N*-isopropylacrylamide) [24] and poly(sodium acrylate) [25], where re-swelling transitions were examined, hydrogen bonds were destroyed in response to the changes in the external conditions, and could not be reformed; furthermore, macroscopic volume changes were a one-way transition. In the case of the present PVA gels, the destruction of microcrystallites by increasing the temperature was reported earlier [16,17]. It was also a one-way transition, and re-formation of microcrystallites was not detected, although the swelling ratio decreased slightly when the temperature decreased. In the present case, the gels returned to a state close to the collapsed phase upon the re-formation of microcrystallites, which were detected for the first time. This evidence suggests that there are two kinds of microcrystallites: stable and unstable, i.e., reversible or irreversible changes in response to the change in solvent composition.

The other important observation is that d/d_0 changed reversibly with a large hysteresis after the second decreasing process. The swelling behavior was reversible in the sense that d/d_0 at high and low acetone concentration was the same in subsequent cycles, suggesting that the destruction and re-formation of microcrystallites that survived the first cycle were reversible in response to a change in solvent concentration. The large hysteresis is attributed to the formation of microcrystallites in the system; once the microcrystallites are destroyed during the decreasing process, subsequent increasing does not cause them to form until a much higher acetone concentration is used.

Finally, it should be noted that the destruction and re-formation of microcrystallites was detected in the present system, which resulted in the hysteresis phenomenon. Hysteresis can be found in many materials systems, such as ferromagnetic, ferroelectric, and viscoelastic materials. Comparing with those phenomena, the present hysteresis was observed in the equilibrium state, which is rate-independent, similar to magnetic hysteresis loops. In hydrogel systems, on the other hand, this phenomenon is reminiscent of multiple phase transitions [26], which are characterized by discontinuous and reversible changes in swelling ratios with large hysteresis, induced by a competitive balance between hydrogen bonds and hydrophobic interactions. Considering the hypothesis of universality of the phase transition of chemical gels, it might be possible to realize a reversible and discontinuous volume change in the present system. For that purpose, it would be necessary to adjust the material or environmental parameters to induce a positive osmotic pressure, and increase the swelling pressure in the good solvent relative to the destruction force of the microcrystallites. This remains a challenging topic in this field.

3. Conclusions

Chemical PVA gels were prepared by crosslinking PVA slightly with glutaraldehyde, washed with a large amount of pure water, and dehydrated fully in air at room temperature to introduce physical crosslinking. Both irreversible and reversible swelling behavior and hysteresis were observed in these PVA gels.

The gel reached its collapsed state in a pure organic solvent by a multiple-sample test, but did not reach the same state in a single-sample test. The equilibrium swelling ratios in a DMSO/acetone mixed solution were measured at room temperature to examine the relation between macroscopic swelling ratio and microscopic network structure. The swollen gel in DMSO could shrink in acetone, but did not return to the initial collapsed state after the gel had experienced the swollen state. This irreversible swelling behavior was attributed to the destruction of the microcrystallites formed during the initial dehydration after gelation. However, the swelling ratio changed continuously and reversibly, with a large hysteresis at around a specific molar ratio of DMSO/acetone of 60/40. This was consistent with the change in the microcrystallites observed by FT-IR measurements; microcrystallites were destroyed in DMSO and re-formed in acetone. This irreversible swelling behavior could not be observed in physical gels, although the sizes and amounts of microcrystallites were similar to those of the chemical gels. These observations were attributed to the irreversible destruction of weak microcrystallites, and to the reversible destruction and re-formation change of the strong microcrystallites.

Finally, it was concluded that the destruction and re-formation of microcrystallites in chemical PVA gels were reversibly controlled by concentration changes in the solvent, which was assisted by the chemical crosslinking. It is important to note that the present findings can be confirmed by other measurements, using XRD, DSC, and other techniques, which merit further research.

4. Experimental

4.1. Sample Preparation

Chemically crosslinked PVA gels with high degrees of polymerization and hydrolysis were obtained by crosslinking PVA solutions with GA. The PVA powder was supplied by Kuraray Co., Ltd. (Cat. No. PVA117, Tokyo, Japan), and was used without further purification. The average degree of polymerization was 1700, and the average degree of hydrolysis was between 98 and 99 mol %. The PVA powder was dissolved in deionized and distilled water at 90 °C or higher for more than 2 h. Gels were prepared in two shapes: thin cylindrical gels for the measurement of swelling ratio were prepared in glass microcapillaries, with an inner diameter of 1.44 mm and inner volume of 50 μL, and thin plate gels for FT-IR measurements were prepared between two slides with a thin spacer (thickness: about 2 mm). The concentration of the monomer units of PVA was fixed at ca. 2100 mM, and the proportion of GA (Wako Pure Chemical Industries, Ltd. (Tokyo, Japan), 25% aqueous solution) to the total concentration of monomer units of PVA plus the monomer of GA was 0.5 mol %. It is noteworthy that the degree of crosslinking of 0.5 mol % GA corresponds to a chain length of about 100 monomer units of PVA.

After mixing GA in dissolved PVA solution, 0.3 mL of 1 M HCl was added to the mixture as a catalyst. The solution was placed in a furnace at 30 °C for 48 h. After gelation, the cylindrical and plate gels were removed from the microcapillaries and mold, and subsequently washed in pure water (the water volume was 1000 times larger than the gel volume) for 48 h, to wash away any residual chemicals or uncrosslinked polymers from the polymer networks. Then, the gels were again placed in an oven at 30 °C for 48 h to dry.

4.2. Measurements of Swelling Ratio and FT-IR

The equilibrium diameter of a cylindrical gel, *d*, was measured by phase contrast microscopy. A sample was placed in respective solvents and left for more than 48 h, typically 96 h, which was sufficient for the gel to reach its equilibrium state. The swelling degree of the gel is represented by the swelling ratio, d/d_0, where d_0 is the gelation diameter (the inner diameter of the glass capillary, 1.44 mm).

To confirm the structural details of the gels, FT-IR measurements using the attenuated total reflectance (ATR) technique was performed on the dehydrated plate gels at room temperature. We used an FT-IR spectrophotometer (Jasco: FT/IR-610, Tokyo, Japan) equipped with an ATR (attenuated total reflectance) attachment with a ZnSe crystal. An appropriate amount of the gel was placed on the ZnSe crystal, and the FT-IR spectra were then recorded at room temperature.

Author Contributions: K.K. and A.S. conceived and designed the experiments; K.K. and S.U. performed the experiments and analyzed the data; Y.H. optimized the method for preparing chemically crosslinked poly(vinyl alcohol) gels and supported the analysis of the FT-IR spectra.

Acknowledgments: We would like to express our sincere gratitude to Kuraray Co., Ltd. (Tokyo, Japan) for supplying PVA powders. This work was supported in part by Japan Society for the Promotion of Science: JSPS KAKENHI grant numbers JP23000011 and JP16H03170.

Conflicts of Interest: The authors declare no conflict of interest.

References

1. Tanaka, T. Gels. *Sci. Am.* **1981**, *244*, 124–136. [CrossRef] [PubMed]
2. Oka, M.; Ushio, K.; Kumar, P.; Ikeuchi, K.; Hyon, S.H.; Nakamura, T.; Fujita, H. Development of artificial cartilage. *Proc. Inst. Mech. Eng. H J. Eng. Med.* **2000**, *214*, 59–68. [CrossRef] [PubMed]
3. Kobayashi, M.; Chang, Y.S.; Oka, M. A two year in vivo study of polyvinyl alcohol-hydrogel (PVA-H) artificial meniscus. *Biomaterials* **2005**, *26*, 3243–3248. [CrossRef] [PubMed]
4. Baker, M.I.; Walsh, S.P.; Schwartz, Z.; Boyan, B.D. A review of polyvinyl alcohol and its uses in cartilage and orthopedic applications. *J. Biomed. Mater. Res. B* **2012**, *100*, 1451–1457. [CrossRef] [PubMed]
5. Hassan, C.M.; Peppas, N.A. Structure and applications of poly(vinyl alcohol) hydrogels produced by conventional crosslinking or by freezing/thawing methods. *Adv. Polym. Sci.* **2000**, *153*, 37–65.
6. Nambu, M. Japanese Patent Kokai. Japan Patent 59/56446, 1984.
7. Lozinsky, V.I. Cryotropic gelation of poly(vinyl alcohol). *Russ. Chem. Rev.* **1998**, *67*, 573–586. [CrossRef]
8. Otsuka, E.; Suzuki, A. A simple method to obtain a swollen PVA gel crosslinked by hydrogen bonds. *J. Appl. Polym. Sci.* **2009**, *114*, 10–16. [CrossRef]
9. Otsuka, E.; Suzuki, A. Swelling properties of physically cross-linked PVA gels prepared by a cast-drying method. In *Gels: Structures, Properties, and Functions*; Springer: Berlin/Heidelberg, Germany, 2009; pp. 121–126.
10. Otsuka, E.; Sugiyama, M.; Suzuki, A. Network microstructure of PVA cast gels observed by SAXS measurements. *J. Phys. Conf. Ser.* **2010**, *247*, 012043. [CrossRef]
11. Wang, B.; Mukataka, S.; Kodama, M.; Kokufuta, E. Viscometric and light scattering studies on microgel formation by γ-ray irradiation to aqueous oxygen-free solutions of poly(vinyl alcohol). *Langmuir* **1997**, *13*, 6108–6114. [CrossRef]

12. Yoshii, F.; Makuuchi, K.; Darwis, D.; Iriawan, T.; Razzak, M.T.; Rosiak, J.M. Heat resistance poly(vinyl alcohol) hydrogel. *Radiat. Phys. Chem.* **1995**, *46*, 169–174. [CrossRef]
13. Praptowidodo, V.S. Influence of swelling on water transport through PVA-based membrane. *J. Mol. Struct.* **2005**, *739*, 207–212. [CrossRef]
14. Dai, W.S.; Barbari, T.A. Hydrogel membranes with mesh size asymmetry based on the gradient crosslinking of poly(vinyl alcohol). *J. Membr. Sci.* **1999**, *156*, 67–79. [CrossRef]
15. Kudo, S.; Otsuka, E.; Suzuki, A. Swelling behavior of chemically crosslinked PVA gels in mixed solvents. *J. Polym. Sci. B* **2010**, *48*, 1978–1986. [CrossRef]
16. Otsuka, E.; Kudo, S.; Sugiyama, M.; Suzuki, A. Effects of microcrystallites on swelling behavior in chemically crosslinked poly(vinyl alcohol) gels. *J. Polym. Sci. B* **2011**, *49*, 96–102. [CrossRef]
17. Otsuka, E.; Sugiyama, M.; Suzuki, A. Formation and destruction of physical crosslinks by mild treatments in chemically crosslinked poly(vinyl alcohol) gels. *Polym. Bull.* **2011**, *67*, 1215–1226. [CrossRef]
18. Hirokawa, Y.; Tanaka, T. Volume phase transition in a non-ionic gel. *J. Chem. Phys.* **1984**, *81*, 6379–6380. [CrossRef]
19. Shinyashiki, N.; Shimomura, M.; Ushiyama, T.; Miyagawa, T.; Yagihara, S. Dynamics of Water in Partially Crystallized Polymer/Water Mixtures Studied by Dielectric Spectroscopy. *J. Phys. Chem. B* **2007**, *111*, 10079–10087. [CrossRef] [PubMed]
20. Nkhwa, S.; Lauriaga, K.F.; Kemal, E.; Deb, S. Poly(vinyl alcohol): Physical Approaches to Designing Biomaterials for Biomedical Applications. *Conf. Pap. Sci.* **2014**, *2014*, 403472. [CrossRef]
21. Miya, M.; Iwamoto, R.; Mima, S. FT-IR study of intermolecular interactions in polymer blends. *J. Polym. Sci. B* **1984**, *22*, 1149–1151. [CrossRef]
22. Mallapragada, S.K.; Peppas, N.A. Dissolution mechanism of semicrystalline poly(vinyl alcohol) in water. *J. Polym. Sci. B* **1996**, *34*, 1339–1346. [CrossRef]
23. Lee, J.; Lee, K.J.; Jang, J. Effect of silica nanofillers on isothermal crystallization of poly(vinyl alcohol): In situ ATR-FTIR study. *Polym. Test.* **2008**, *27*, 360–367. [CrossRef]
24. Hirashima, Y.; Suzuki, A. Roles of Hydrogen Bonding on the Volume Phase Transition of Ionized Poly(N-isopropylacrylamide) Gels. *J. Phys. Soc. Jpn.* **2004**, *73*, 404–411. [CrossRef]
25. Sato, H.; Hirashima, Y.; Suzuki, A. Reswelling transition of poly(sodium acrylate) gels due to destruction of hydrogen bonds observed by ATR FTIR spectroscopy. *J. Appl. Polym. Sci.* **2007**, *105*, 3809–3816. [CrossRef]
26. Annaka, M.; Tanaka, T. Multiple phases of polymer gels. *Nature* **1992**, *355*, 430–432. [CrossRef]

gels

MDPI

Article

Dynamic Behaviors of Solvent Molecules Restricted in Poly (Acryl Amide) Gels Analyzed by Dielectric and Diffusion NMR Spectroscopy

Hironobu Saito [1], Shunpei Kato [2], Keisuke Matsumoto [2], Yuya Umino [2], Rio Kita [2], Naoki Shinyashiki [2], Shin Yagihara [2,*], Minoru Fukuzaki [3] and Masayuki Tokita [4]

[1]   Graduate School of Science and Technology, Tokai University, Kanagawa 259-1292, Japan; hilonobu.saito@gmail.com
[2]   Graduate School of Science, Tokai University, 4-1-1 Kitakaname, Hiratsuka-shi, Kanagawa 259-1292, Japan; chikyuunimoyasasiku@gmail.com (S.K.); se2gekka@gmail.com (K.M.); yuno_estrade@ezweb.ne.jp (Y.U.); rkita@keyaki.cc.u-tokai.ac.jp (R.K.); naoki-ko@keyaki.cc.u-tokai.ac.jp (N.S.)
[3]   Liberal Arts Education Center in Kumamoto, Tokai University, Toroku 9-1-1, Higashi-ku, Kumamoto-shi, Kumamoto 862-0970, Japan; mfukuzaki@tokai-u.jp
[4]   Graduate School of Science, Kyushu University, 744, Motooka, Nishi-ku, Fukuoka 819-0395, Japan; tokita@phys.kyushu-u.ac.jp
*   Correspondence: yagihara@keyaki.cc.u-tokai.ac.jp; Tel.: +81-463-58-1211

Received: 7 April 2018; Accepted: 14 June 2018; Published: 22 June 2018

Abstract: Dynamics of solvent molecules restricted in poly (acryl amide) gels immersed in solvent mixtures of acetone–, 1,4-dioxane–, and dimethyl sulfoxide–water were analyzed by the time domain reflectometry method of dielectric spectroscopy and the pulse field gradient method of nuclear magnetic resonance. Restrictions of dynamic behaviors of solvent molecules were evaluated from relaxation parameters such as the relaxation time, its distribution parameter, and the relaxation strength obtained by dielectric measurements, and similar behaviors with polymer concentration dependences for the solutions were obtained except for the high polymer concentration in collapsed gels. Scaling analyses for the relaxation time and diffusion coefficient respectively normalized by those for bulk solvent suggested that the scaling exponent determined from the scaling variable defined as a ratio of the size of solvent molecule to mesh size of polymer networks were three and unity, respectively, except for collapsed gels. The difference in these components reflects characteristic molecular interactions in the rotational and translational diffusions, and offered a physical picture of the restriction of solvent dynamics. A universal treatment of slow dynamics due to the restriction from polymer chains suggests a new methodology of characterization of water structures.

Keywords: poly (acryl amide) gel; time domain reflectometry (TDR) of dielectric spectroscopy; pulse field gradient spin echo method of nuclear magnetic resonance (PFG-NMR); scaling analysis; fractal analysis

1. Introduction

Polymer gels, i.e., crosslinked polymer networks with solvent have been extensively applied in diverse fields such as medical, environment, and food science and industries because of the characteristic behaviors. One of the most typical and characteristic behaviors of polymer gels is a volume phase transition with which the late Professor Tanaka has first indicated that fundamental physics of polymer gels, leading to expanded usage in applied science. Poly (acryl amide) (PAAm) gels swell and shrink sensitively with a change in environment, such as temperature, electric field, composition of solvent, ion concentration, pH, etc. [1–6]. However, there still exist some ambiguities in physical properties of the polymer gel related to the volume phase transition, especially from a view point of molecular interactions among the solvent molecules and polymer chains.

One of the authors (M.T.) has reported that diffusion coefficients of probe molecules obtained by the pulsed field gradient spin echo method of nuclear magnetic resonance (PFG-NMR) [7,8] showed dynamic behaviors of probe molecules restricted in polymer gels [9–11]. The mesh size of the network could be evaluated from the restriction of various probe molecules with different sizes. This methodology can be an effective tool for the evaluation of gels of not only synthetic but also natural polymers [12,13]. However, this method generally requires some probe molecules, since water protons cannot be used with the recent high-resolution equipment because of oscillation dumping. There has not been a lot of techniques to evaluate the volume phase transition without addition of probe molecules.

We have investigated the dynamic behavior of solvent molecules restricted in polymer networks of gels [14,15] by using the time domain reflectometry (TDR) method [16–18] of microwave dielectric spectroscopy. The TDR method is capable of observing relaxation processes in the GHz frequency region including the most important frequencies for liquid molecules at ordinary temperatures and those slow dynamics with restrictions in polymer gels. Furthermore, open-end coaxial electrodes, which can be attached to the surface of gel and solid materials for dielectric measurements with the fringing field, are easily used in TDR measurements [14]. Therefore, in the case of dielectric studies on the volume phase transition, the TDR measuring system is not simply a conventional method, and various refinements makes the TDR method the most effective tool for detailed analysis and discussion, especially for cooperative dynamics of mixed solvent.

Recently, Yang and Zhao, et al. [19–21], reported dielectric properties of the temperature-sensitive phase transition of poly (N-isopropylacrylamide) (PNiPAM)-based microgels, for which conventional electrodes are available. The PNiPAM gel is also typical polymer gels expected to be applied for medial usages like microcapsules and drug delivery systems. Relaxation processes observed at frequencies lower than 110 MHz are of interest in chain dynamics, however, solvent dynamics and behaviors are not directly reflected in the frequency region.

In the present work, we examined acetone, 1,4-dioxane, and dimethyl sulfoxide (DMSO) as organic solvents of aqueous mixtures with various compositions, in order to investigate the molecular mechanism bringing abrupt volume changes of PAAm gels with the composition of the organic solvent–water mixtures. PFG-NMR and TDR measurements were performed to compare dynamic behaviors of solvent molecules inside gels with those for bulk liquid mixtures. Various difficulties, such as the contact of TDR electrodes with samples, high frequency measurements of water, and oscillation dumping for water protons in TDR and NMR measuring techniques for the gel study require more creative treatments of the equipment. Furthermore, the complementarily analysis with the characteristic time scales of 1 ms for NMR and 10 ps for TDR measurements offers us more detailed analysis of the fluctuation of the order parameters, such as the relaxation time distributions. The three organic solvents were chosen because of their different characteristics: 1,4-dioxane is a non-polar liquid; acetone and DMSO have a similar chemical structure. Each component of the solvent mixtures respectively interacts with polymer chains with different affinity. The differences in their dynamics reflected the restricted behavior of these solvent molecules and led us to more detailed analysis of the intermolecular interactions among solvent molecules and polymer chains.

2. Results and Discussion

2.1. Swelling Ratio of Gels

Figure 1 shows mole fraction dependences of the swelling ratio, V/V_0, for PAAm gels immersed in aqueous mixtures of acetone, 1,4-dioxane, and DMSO, respectively, with some concentrations raised step by step to reach an equilibrium volume at the target concentration. Here, V was determined from the equilibrium volume of the gel at certain composition of the solvent mixtures and V_0 is the initial volume of the gel prepared in pure water. The initial volume, V_0, was determined when the gels were synthesized at the polymer concentration of about 5.1 wt %. Upon immersing the initial gel in pure water, the gel swelled and reached the equilibrium volume. Then, the value, V/V_0, was larger than unity in water. The result for the acetone–water mixtures shows an abrupt change in the mole fraction region of $0.1 < \chi_{organic} < 0.2$ as reported by Tanaka et al. [1]. A similar result was also obtained for the gel immersed in 1,4-dioxane aqueous solution. On the other hand, the PAAm gel with DMSO aqueous solution showed a similar but a smaller volume change and an opposite volume change appeared in the DMSO-rich region.

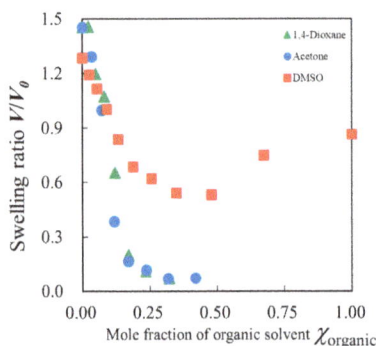

Figure 1. The swelling ratio for poly (acryl amide) (PAAm) gel normalized by the initial value in water, V/V_0, dependent on the mole fraction of organic solvent in the aqueous solution for acetone, 1,4-dioxane, and dimethyl sulfoxide (DMSO).

2.2. PFG-NMR Measurements

Usual application of the PFG-NMR method to gel materials requires probe molecules [9,11,13]. We have reported that liquid structures, including hydrogen bonding networks, reflect larger scales of interactions [22,23]. Thus, protons of water and organic solvent molecules were used to examine the molecular diffusion processes in the present work. Figure 2a shows acetone and DMSO mole fraction, $\chi_{organic}$, dependences of the diffusion coefficients for water molecules of the solvent mixtures inside gels and the bulk solvent obtained from PFG-NMR method.

All plots decreased with increasing mole fraction of organic solvents in the region, $\chi_{organic} < 0.2$. The diffusion coefficient of solvent molecules restricted in the gels were smaller than those in the bulk solvents. However, composition dependences of the diffusion coefficient for DMSO were different from those for acetone. Former dielectric studies on aqueous solutions of organic solvent usually showed changes in composition dependences of relaxation parameters at around the mole fraction of water, 0.83, even if those changes are only apparent [24,25]. Actually, our results on the diffusion coefficient for the present solvent mixtures also apparently indicated changes in the composition dependences as shown at around the mole fraction of the organic solvent, 0.17. The solvent mixtures take characteristic values of the diffusion coefficient in the region, $\chi_{organic} > 0.2$, as shown in Figure 2a.

Figure 2b shows similar results of the restrictions of dynamic behaviors also for the organic solvent molecules. The smaller values of the diffusion coefficient for the organic solvents than those for water molecules mean more restrictions because of the larger size of molecules. It is available even in the case of solvent mixtures without gels. Furthermore, comparing with the size of solvent molecules, the mesh size is generally large enough, since the ratio of the tetrafunctional monomer used for crosslinking was just 1/20 against the bifunctional monomers for chains. Therefore, the restrictions were almost entirely from polymer chains and the effect of shrinking is accompanied with increasing polymer concentrations. The restriction we compared among the organic solvents indicated the solvent size dependency in the diffusion coefficient.

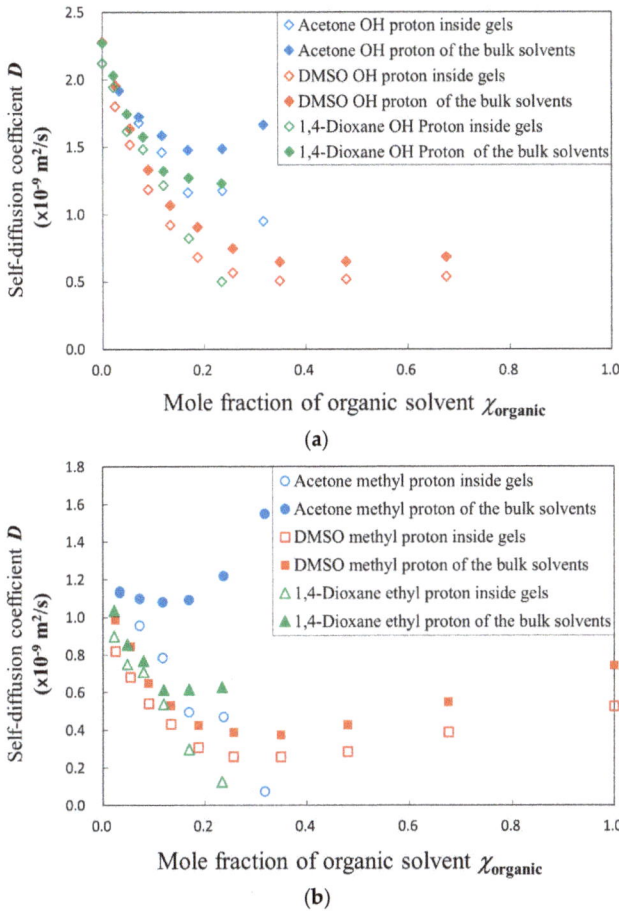

Figure 2. Acetone and DMSO mole fraction dependence of the diffusion coefficient for solvent molecules inside gel or bulk solvent obtained from: (**a**) Proton of water molecules; (**b**) Methyl proton of organic solvent molecules.

In order to focus attention on the restriction from the polymer chain networks in the gels, the diffusion coefficient of solvent molecules in the gels was normalized by those obtained in the bulk solvent mixtures. Figure 3a shows plots of the logarithm of the normalized diffusion coefficient, D_{gel}/D_{sol}, against the composition of solvent mixtures for acetone, 1,4-dioxane, and DMSO aqueous solutions. The normalized diffusion coefficient for acetone and 1,4-dioxane became smaller with shrinking gels, but DMSO did not show similar behavior, since the swelling ratio of gels for DMSO aqueous solutions was too large to restrict solvent dynamics. The same explanation is available for larger values of the normalized diffusion coefficient for water molecules. The composition dependence of the normalized diffusion coefficient for 1,4-dioxane aqueous solution was similar to the acetone aqueous solution as their swelling ratio behaviors were also similar. Figure 3b shows composition dependences of density for the organic solvent aqueous solutions. Characteristic properties of density for each organic solvent's aqueous solution were not directly reflected in the dynamic behaviors, since the dynamics restricted by polymer chain networks were affected more by the swelling ratio.

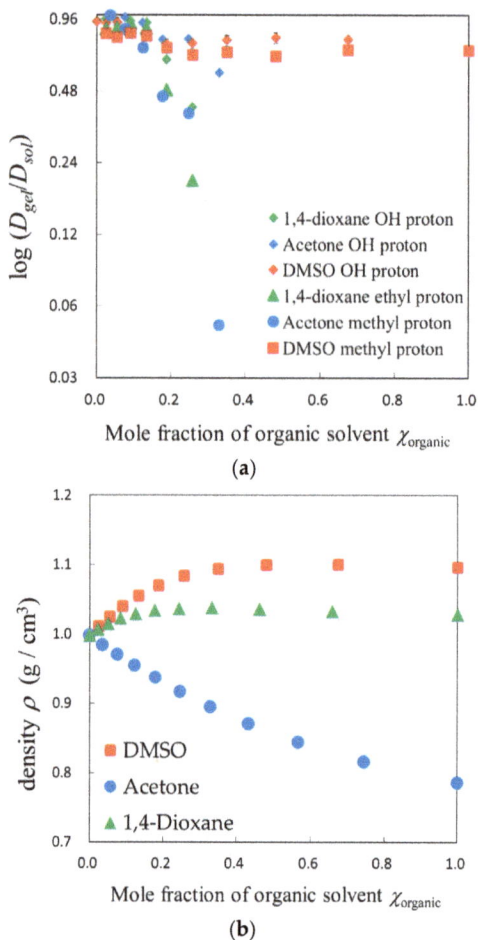

Figure 3. Mole fraction of organic solvent dependences of the diffusion coefficient and density for each solvent mixture: (**a**) The diffusion coefficient of solvent in gels normalized by those in bulk solvent; (**b**) The density of bulk solvent mixtures.

2.3. TDR Meserments

Figures 4 and 5 show dielectric dispersion and absorption curves obtained by TDR measurements for acetone and DMSO aqueous solutions with PAAm gels. The binary mixtures of polar molecules show only one relaxation process, even if these diffusion coefficients show two values. This is also an indication of an averaging effect of the dielectric properties with the large-scale behaviors of hydrogen bonding liquids. Restrictions of the dynamic behaviors appear as a lower frequency shift of the peak frequency and a decrease in the relaxation strength. Differences between the curves for the solvent inside and outside the gel is larger in the water poor region with smaller swelling ratio. Then, comparing Figures 4 and 5, the larger differences between solvent molecules inside and outside the gels shown for acetone aqueous solutions are simply explained from a larger decrease in the swelling ratio. The difference indicating the restriction of molecular dynamics of solvent in the polymer network is characterized by a lower frequency shift of the peak frequency and a smaller relaxation process with increasing polymer concentration with the shrinkage of gels.

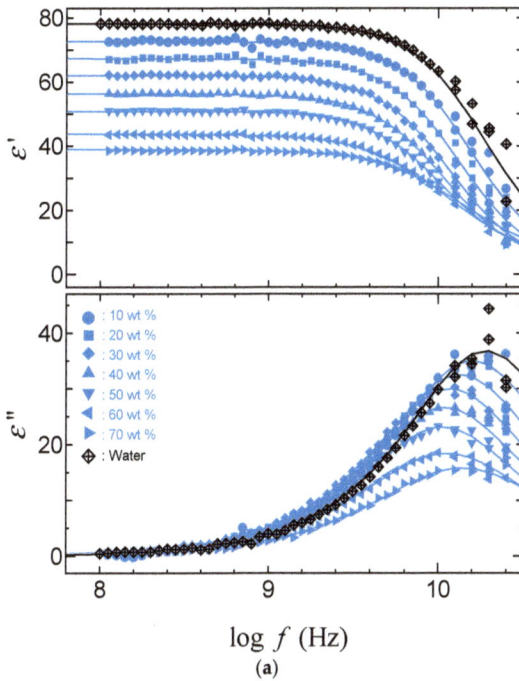

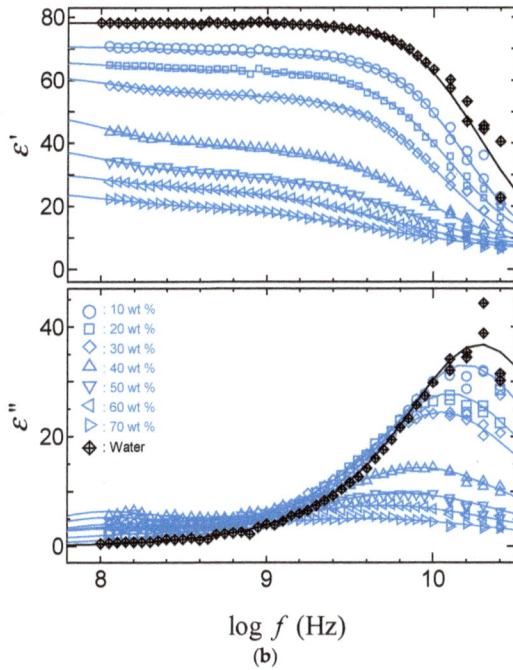

Figure 4. Dielectric dispersion and absorption curves for acetone–water mixtures: (**a**) Outside PAAm gels; (**b**) Inside PAAm gels.

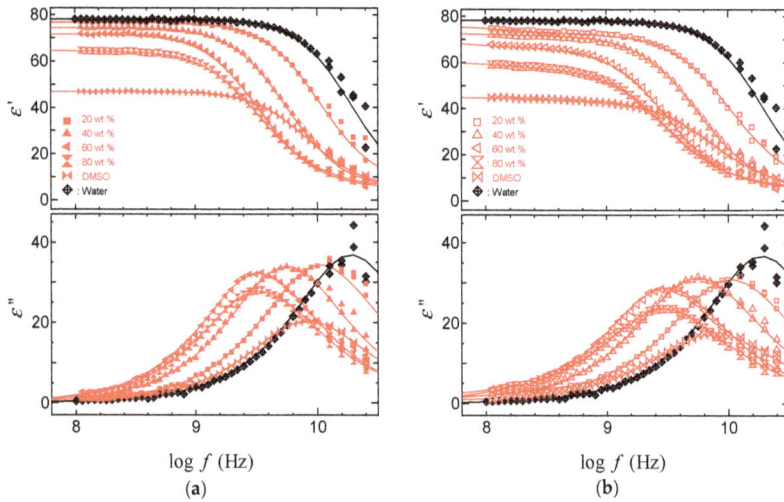

Figure 5. Dielectric dispersion and absorption curves for DMSO–water mixtures: (a) Outside PAAm gels; (b) Inside PAAm gels.

The relaxation parameters were obtained from the fitting procedures to dielectric relaxation data with the following Equation,

$$\varepsilon^* = \varepsilon\prime - j\varepsilon'' = \varepsilon_\infty + \frac{\Delta\varepsilon}{[1 + (j\omega\tau)^\beta]^\alpha} - j\frac{\sigma_{DC}}{\varepsilon_0\omega} \tag{1}$$

where ε' and ε'' are the real and imaginary parts of the complex dielectric constant, ε^*, respectively, j is the imaginary unit, ε_∞ is the limiting high-frequency dielectric constant, $\Delta\varepsilon$ is the relaxation strength, ω is the angular frequency, τ is the relaxation time, α and β $(0 < \alpha, \beta \leq 1)$ is the relaxation time distribution parameter, and σ_{DC} is the dc conductivity, ε_0 is the dielectric constant of vacuum. The relaxation function used for the fitting procedure was Cole–Cole Equation ($\alpha = 1$) [26] for 1,4-dioxane and acetone aqueous solutions, Cole-Davidson Equation ($\beta = 1$) [27] for bulk DMSO aqueous solutions [28,29], and Havriliak–Negami Equation ($0 < \alpha, \beta \leq 1$) [30] for DMSO aqueous solutions inside gels.

Figure 6 shows composition dependences of the relaxation time and the relaxation strength. The larger relaxation time corresponds to the large frequency shift to the lower frequency side. The relaxation strength was smaller for the solvent inside gels because of the decrease in the density of solvent molecules.

Figure 7 shows polymer concentration dependence of the normalized relaxation strength for each solvent mixture. The normalized relaxation strength for aqueous solutions of PAAm and poly (acryl acid) (PAA) were also shown for comparison. DMSO aqueous solutions kept larger values, since the gel did not shrink enough even at low water content. Results for acetone- and 1,4-dioxane-water mixtures show similar dependency of decreasing swelling ratio, but the values of normalized relaxation strength for 1,4-dioxane–water mixtures were larger than those for acetone–water solutions, since the 1,4-dioxane is a non-polar organic solvent. Furthermore, the polymer concentration dependence of the normalized relaxation strengths for acetone aqueous solutions showed similar values to those shown for water in PAA aqueous solution and the same kind of behavior as those for PAA and PAAm aqueous solutions in the polymer concentration region less than 0.6, but the dependency exhibited a different manner in the region above 0.6. This result suggests the existence of another phenomenon in collapsed gels.

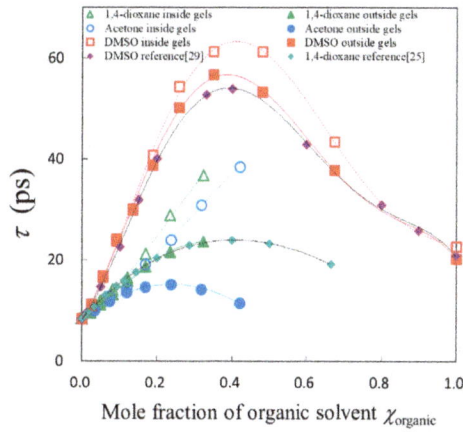

(a)

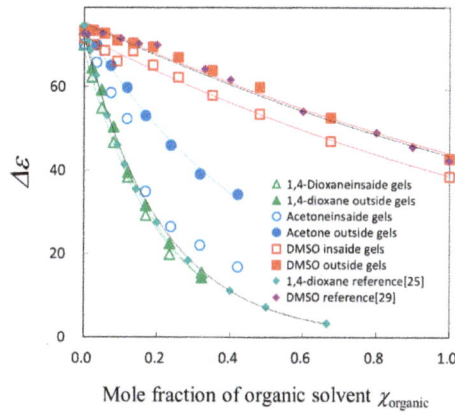

(b)

Figure 6. Composition dependences of the relaxation parameters: (**a**) Relaxation time; (**b**) Relaxation strength.

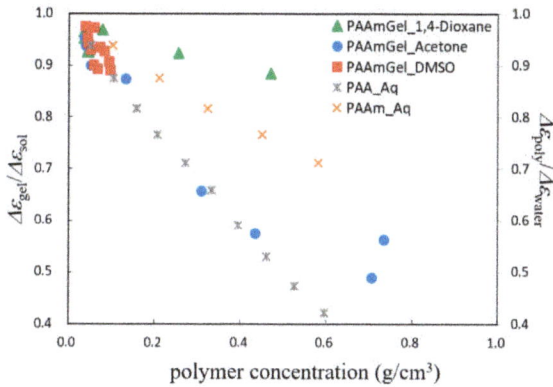

Figure 7. Polymer concentration dependence of the normalized relaxation strength for each solvent mixture.

The relaxation time for solvent mixtures in gels normalized by those outside gels are plotted against the composition of each solvent mixture in Figure 8. Characteristic behaviors were shown for the normalized relaxation time in the mole fraction region of $0.1 < \chi_{organic} < 0.2$, in which the swelling ratio abruptly changed. Aqueous molecular liquids apparently show a discontinuous behavior around a characteristic mole fraction of water, 83%, though Buchner et al. reported that there exists no corresponding characteristic liquid structure [31]. In the mole fraction region of $0.2 < \chi_{organic} < 0.5$, the normalized relaxation time for each solvent mixture seems to increase simply with shrinking gel. The normalized relaxation parameters reflect the swelling ratio well.

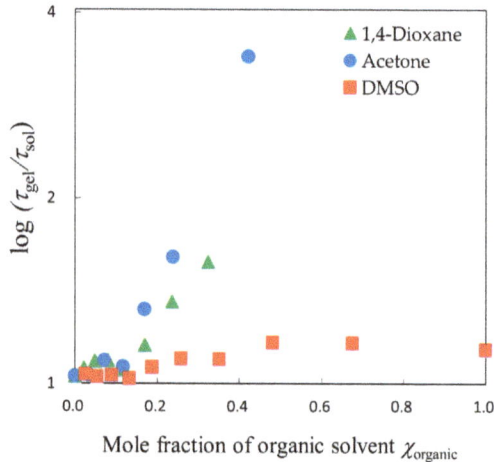

Figure 8. Composition dependences of the relaxation time; log plot of the normalized relaxation time against the mole fraction of organic solvent of the aqueous solutions.

Composition dependence of the normalized relaxation time is shown in Figure 8, the normalized relaxation time reflects much more restrictions from the polymer network through the swelling ratio. Considering that the chain dynamics are slower than those of solvent molecules in the present study, it is reasonable that slow dynamics of solvent molecules are determined by chain dynamics of low mobility polymer networks more than solvent dynamics with high mobility.

The relaxation time for the dipole relaxation process is a characteristic time of molecular dynamics of the rotational diffusion, but it is just a measure of the average value. Then, the distribution of the relaxation time is necessary for more exact analysis and detailed discussion. Figure 9a shows composition dependence of the relaxation time distribution parameters, α and β, defined by Equation (1), and the β values are plotted against the polymer concentration in Figure 9b. The parameters, α and β, are related to asymmetric and symmetric broadening of the relaxation curve [31]. The physical meanings of the parameters, α and β, are molecular interactions similar to the chain connectivity of polymers and fractal fluctuations of density, respectively [32]. Then, parameters α and β can be respectively treated even in the case of relaxation processes described by the Havriliak–Negami Equation. Figure 9b shows similar broadening for three organic solvent mixtures with increasing restriction of shrinking gels in the concentration region between 0.1 and 0.6, except for the characteristic behaviors shown at concentration below 0.1. In the high concentration region above 0.6, no clear concentration dependence is shown. This result suggests a possible explanation in which the composition of the solvent mixtures inside the gel is different from that outside the gel. The organic solvent concentration in the gel is considered to be lower than that expected at the same composition of solvent mixtures existing outside gel, since PAAm chains are hydrophilic.

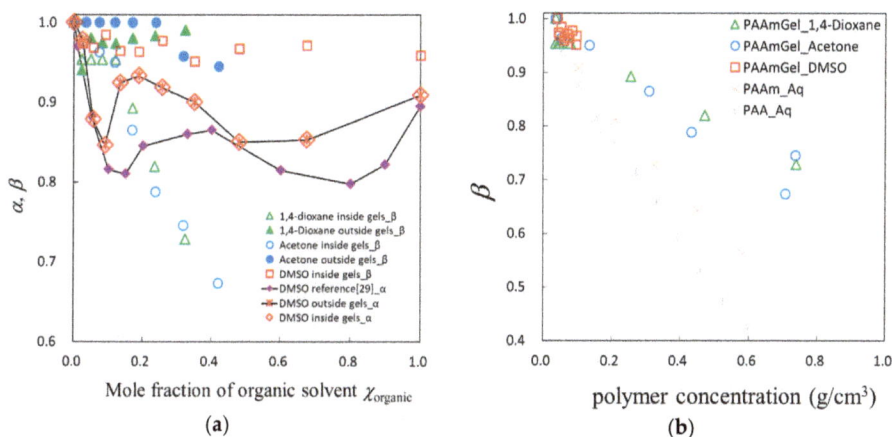

Figure 9. Behaviors of the relaxation time distribution parameter: (**a**) Composition dependence of the relaxation time distribution parameters, α and β, for solvent mixtures inside or outside the gels; (**b**) Polymer concentration dependence of the relaxation time distribution parameter, β, for solvent mixtures inside.

Logarithm of the normalized relaxation time, τ_{gel}/τ_{sol}, and the normalized diffusion coefficient, D_{gel}/D_{sol}, are plotted against polymer concentration in Figure 10a,b. Polymer concentration dependence of the relaxation time and the diffusion coefficient show reasonable tendencies, respectively, reflecting restrictions from shrinking polymer networks except for the low polymer concentration region less than 0.1 g/cm³, in which characteristic behaviors are shown.

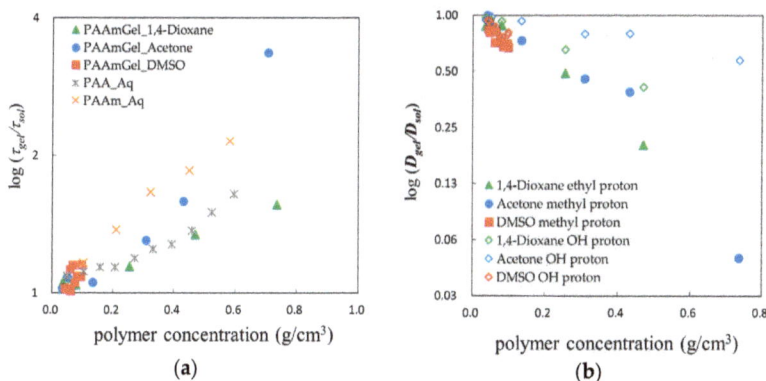

Figure 10. Polymer concentration dependences of the relaxation time and diffusion coefficient for solvent molecules restricted in polymer chain networks: (**a**) Logarithm of the normalized relaxation time; (**b**) Logarithm of the normalized diffusion coefficient.

2.4. Scaling Concepts

The interactions among mixed solvents and polymer chains in gels clearly showed characteristic features of those dynamic properties. Though the dynamic properties of solvent mixtures in gels were normalized by those in bulk solvent mixtures, each system still shows characteristic behavior. Scaling concepts were used to examine the universal property of the restrictions.

Solvent dynamics restricted by shrinking gels have been analyzed by scaling law, especially for translational diffusion constants obtained from PFG-NMR measurements for probe or solvent molecules [9,10,33]. To investigate the molecular dynamics of solvent restricted in polymer networks, the scaling variable expressed by a ratio of sizes of the solvent molecule and the mesh size of the polymer networks can suggest a physical picture of the molecular mechanism, as

$$x = R/\xi \tag{2}$$

where the scaling variable, x, is expressed as a ratio of sizes of the solvent molecule, R, and the mesh size, ξ, of the polymer chain networks. Following power law relationship, R and ξ are rewritten by the molecular weight of solvent molecules, M, and polymer concentration, C_p [34] as

$$R \propto M^{\frac{1}{3}}, \; \xi \propto C_p^{\frac{-3}{4}} \tag{3}$$

Using Equation (3) and suitable exponents, the normalized relaxation time, τ_{gel}/τ_{sol}, and the diffusion coefficient, D_{gel}/D_{sol}, were finally expressed by

$$\frac{\tau_{gel}}{\tau_{sol}} \propto f\left(x^3\right) = f\left\{\left(\frac{R}{\xi}\right)^3\right\} = exp\left\{\left(M^{\frac{1}{3}}C_p^{\frac{3}{4}}\right)^3\right\} \tag{4}$$

and

$$\frac{D_{gel}}{D_{sol}} \propto f\left(x^{-1}\right) = f\left\{\left(\frac{R}{\xi}\right)^{-1}\right\} = exp\left\{\left(M^{\frac{1}{3}}C_p^{\frac{3}{4}}\right)^{-1}\right\} \tag{5}$$

Here, the scaling exponents were three and unity, respectively, for Equations (4) and (5). The M values used for the scaling analysis with Equation (5) were determined as those for the probe of solvent molecules. On the other hand, the M values used for the scaling analysis with Equation (4) were determined as averages following the composition of water and organic solvent molecules, since only a single cooperative relaxation process is observed for the solvent mixtures. Other values of scaling exponent could not represent straight lines. These analyses of the scaling law suggest that the scaling variable x (=$M^{1/3}C_p^{3/4}$) is available for the normalized diffusion coefficient D_{gel}/D_{sol} obtained by the PFG-NMR method and x^3 (=$MC_p^{9/4}$) for the normalized relaxation time, τ_{gel}/τ_{sol} obtained by the TDR method.

Figure 11a,b finally shows linear relationships if the plot for the most collapsed gel for the acetone aqueous solution is neglected. Error bars shown in Figure 11 express the accuracy reflecting volume, NMR, and dielectric measurements. These obvious differences for the plot for the most collapsed gel for acetone aqueous solution were also shown in Figures 7 and 9b. Considering the effect of the existence of a small amount of water molecules remaining in the most collapsed gel with the hydrophilic properties of PAAm, the plot shifts to the larger value of D_{gel}/D_{sol} and the smaller value of τ_{gel}/τ_{sol}, respectively, in Figure 11a,b. These compensations tend to return those plots to straight lines.

The result of different scaling exponents obtained for Equations (4) and (5) is supposed to reflect the difference in the physical meanings of the relaxation time and the diffusion coefficient. Both physical properties related to diffusive dynamics of molecules in the medium are expressed with friction described by the Stokes law, as

$$D = \frac{kT}{6\pi\eta r} \tag{6}$$

and

$$\tau_{rot} = \frac{V\eta}{kT} \tag{7}$$

here k is the Boltzmann constant, T is the absolute temperature, r and V are the radius and the volume of molecules, respectively. Equation (7) implies that the relaxation time is determined by the intermolecular interactions expressed by the ratio of volumes for the solvent molecule and the mesh of polymer network. Therefore, this expression for the relaxation time seems to be reasonable, since typical expression of dynamic behaviors are often treated with the free volume theory.

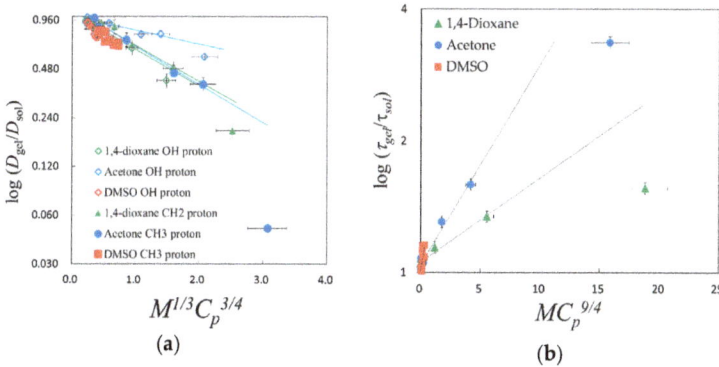

Figure 11. This is a figure, Schemes follow the same formatting. If there are multiple panels, they should be listed as: (**a**) Description of what is contained in the first panel; (**b**) Description of what is contained in the second panel. Figures should be placed in the main text near to the first time they are cited. A caption on a single line should be centered.

2.5. Fractal Analysis with τ–β Diagram

Recently, we have examined fractal analysis for evaluation of the water structure [35–37]. Ryabov et al. expressed the relationship between the Cole–Cole relaxation time distribution parameter and the relaxation time [38,39] as

$$\beta = \frac{d_G}{2} \frac{\ln(\tau \omega_s)}{\ln(\tau/\tau_0)} \tag{8}$$

where τ_0 is the cutoff time of the scaling in time domain, d_G is the fractal dimension of the point set where relaxing units are interacting with the statistical reservoir,

$$\omega_s = 2 d_E G^{2/d_G} D_s / R_0^2 \tag{9}$$

is the characteristic frequency of the self-diffusion process where d_E is the Euclidean dimension, D_s is the self-diffusion coefficient, R_0 is the cutoff size of the scaling in the space, and G is a geometrical coefficient approximately equal to unity. This analysis requires a fractal dimension to combine both the average value of characteristic time of dynamics (the relaxation time) and the fluctuation (Cole–Cole relaxation time distribution parameter), and it is not necessary to know any exact values of the water content. Finally, we know how water molecules aggregate and disperse in materials from the fractal analysis.

Equation (8) was examined for the GHz frequency process observed in the present work. Generally, the fractal analysis requires plotting of the Cole–Cole relaxation time distribution parameter against the logarithm of the relaxation time, and obtained hyperbolic curve is analyzed. In the present work, the normalized relaxation time, τ_{gel}/τ_{sol}, was used for the analysis, since slow dynamics due to restriction from shrinkage of the polymer network were treated. Figure 12 shows the τ–β diagram for the present PAAm gels with organic solvent aqueous solutions. The plots obtained for the solvent molecules restricted in polymer chains show hyperbolic curves for aqueous solutions. Usually, plots for gels are located in the region lower than those for solutions in the Figure, and the water structures in gels are more heterogeneous than those for solutions. In the present work, however, the curves obtained for organic solvent aqueous solutions restricted in PAAm gels take similar shape and occur in a slightly lower region compared to curves for aqueous solutions of PAAm and PAA. The plot for the collapsed gel occurs in the upper right region because of the under-estimation of τ_{sol} for the remaining water molecules. This tendency means that the mesh size of the polymer network cannot be more homogeneous than the polymer chains in the solutions, and this result follows the characteristic behaviors of gel, solution, and dispersion systems which we have obtained in recent works [35,37,40].

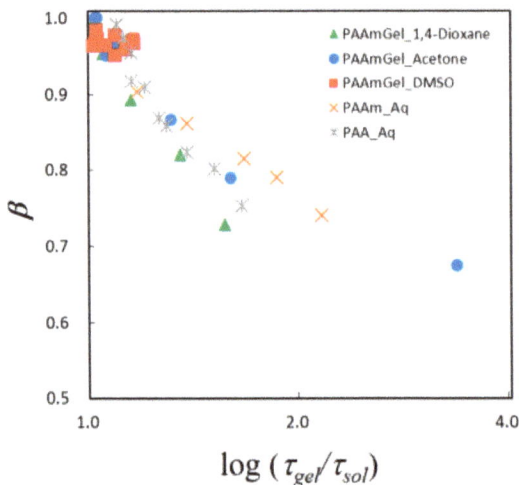

Figure 12. Plot of the logarithm of the relaxation time distribution parameter of Cole–Cole function against the normalized relaxation time, τ_{gel}/τ_{sol}.

The fractal analysis performed for the GHz frequency process with 10 ps time scale obtained from TDR measurements cannot be treated in the same manner as NMR measurements with 1 ms time scale. The dynamic properties of cooperative interactions of hydrogen bonding networks treated by TDR measurements were not reflected in larger scale observation of diffusion coefficients caused by the averaging effect [37]. The fractal analysis used in dielectric study is remarkably useful for evaluation of water structures, especially for investigating how water molecules are coagulated and dispersed.

3. Conclusions

Restrictions of solvent molecule dynamics in PAAm gels were analyzed by TDR and PFG-NMR measurements for mixed solvents of acetone–, 1,4-dioxane–, and DMSO–water. The restrictions could be expressed by the scaling law with a scaling variable of the ratio between the size of solvent molecules and the mesh size of the PAAm network. Suitable exponents were determined as unity and three for the diffusion coefficient and the relaxation time, respectively. The fractal analysis suggests that the water structure of polymer networks cannot be more homogeneous than those in polymer solutions.

4. Materials and Methods

4.1. Materials

The preparation of the poly (acrylamide) (PAAm) gel was following Tanaka et al. [1]. Acrylamide (5 g), *N,N'*-methylene-bis-acrylamide (0.133 g), ammonium persulfate (40 mg), and *N,N, N,N*-tetra-methylethylene-diamin (TEMED) (240 µL) were dissolved in distilled and deionized water (milli-Q system: Merck Millipore Japan Co., Ltd., Tokyo, Japan) to a final volume of 100 mL. The solution was poured into glass tubes with a diameter of 7 mm. It appears to take approximately 2 h for gelation at room temperature. The gel was cut in the 10 mm column and immersed in deionized water for 3 days to wash away residual acrylamide, bis-acrylamide, ammonium persulfate, and TEMED. The volume of the gels decreased with increasing the organic solvent composition and volume phase transitions were observed. The aqueous solution of monomer, initiator, and reaction accelerator was prepared and was put into a glass tube (length: 50 mm, diameter: 8 mm) and was kept for 2 h. The gel was cut in the 10 mm column and immersed in pure water for 2 days to wash away residual acrylamide, bis-acrylamide, and other impurities. The gels were immersed in the solvent of various compositions.

Acetone, 1,4-dioxane, and dimethyl sulfoxide (DMSO) were used as organic solvent to decrease the volume of the gels. The gels were placed in acetone–water and 1,4-dioxane–water mixtures with those concentrations from 0 to 70 wt % at 10 wt% intervals. In addition, DMSO–water mixtures with those concentrations from 0 to 100 wt % at same intervals.

In NMR measurements, the samples were cut out by cover glass and put into the aspirator tube with outer diameter: 2.0 mm and the inner diameter: 1.4 mm. In the case of acetone and 1,4-dioxane aqueous solutions, the size of samples obtained for more than 70 wt % organic solvent were too small to make measurements.

4.2. Experimental Methods

4.2.1. Volume, Density, and Viscosity

The volume of the gel was calculated from diameter and length determined by caliper. The diameter and length were averages of 5 times measurements. Density measurements for 10 wt % organic solvent–water mixtures were performed at 25 °C by density meter DMA48 (Anton Paar, Tokyo, Japan).

4.2.2. NMR Measurements

Nuclear magnetic resonance (NMR) was used to determine the diffusion coefficient. The experiments were performed on a nuclear magnetic resonance spectrometer (EX-90, JEOL, Tokyo, Japan), which was equipped with a pulsed field gradient spin echo (PFG-SE). Figure 13a shows the pulse sequence used for PFG-SE NMR measurements. The principles of the PFG-SE technique have already been reported in detail [32]. For the calibration of the gradient magnetic field strength, the diffusion coefficient of Reference [41] was used. The temperature was controlled to 25.0 ± 0.2 °C. The diffusion coefficients of the probe molecules are determined from the intensity of spin echo signal [7]. The intensity of the spin echo signal, A, in the presence of the field gradient pulses is expressed as follows:

$$A = A_0 exp\left[-\gamma^2\delta^2 G^2\left(\Delta - \frac{\delta}{3}\right)D\right] \tag{10}$$

here, A_0 is the echo amplitude in absence of the field gradient pulses, γ is the magnetogyric ratio of the observed nucleus, δ is the duration of the field gradient pulse, 0.1~2.0 ms. G is the intensity of pulse field gradient, 87 gauss/cm. Δ is the time interval, 20 ms, between the leading edges of the field gradient pulses, and D is the diffusion coefficient of the probe molecule, respectively.

Figure 13b shows an example of spin echo signal attenuation of two peaks, respectively, for water and DMSO protons obtained at every 0.4 ms of δ between 0 and 2.0 ms. Each D value was obtained from fitting procedures of the attenuation of normalized amplitude obtained at every 0.1 ms of δ between 0.3–1.3 ms with Equation (10).

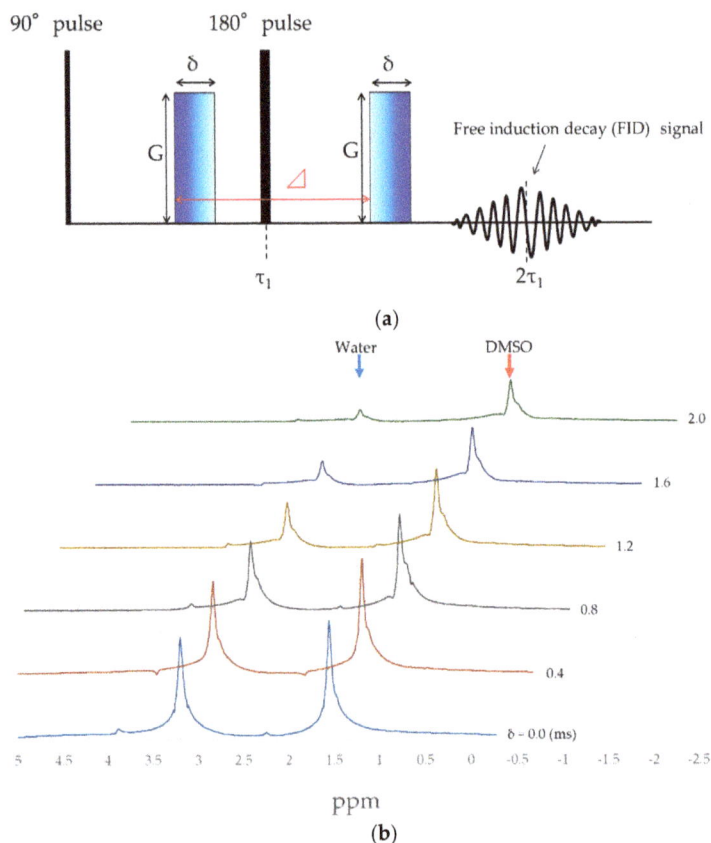

Figure 13. (a) The pulse sequence for the spin-echo method for measurement of diffusion coefficient by pulsed field gradient. Block pulse written by a diagonal gradation is a magnetic gradient; **(b)** The spin echo signal attenuation of 1 H NMR spectra for 60 wt % DMSO aqueous solutions inside gels by varying field gradient pulse duration, δ.

4.2.3. Dielectric Measurements

TDR measurements were performed by digitizing oscilloscope (HP54120B, Agilent Technology, Tokyo, Japan) and Four Channel Test Set (HP54124A, Agilent Technology) with a homemade open-end coaxial electrode with an outer diameter of 2.2 mm. Contact of the open-end of the electrodes to the surface of the gels offers a fringing field penetrating inside the gel and practical dielectric measurements. Dielectric measurements of the solution were performed for solvent outside the gel. Applied voltage was 200 mV and time ranges used were 50, 100, 200, 500, and 1000 ps/div. Temperature was controlled by a homemade temperature jacket at 25.0 ± 0.5 °C.

Author Contributions: H.S. and S.K. performed experiments, analysis, and wrote the paper; K.M. and Y.U. performed the experiments and analysis; R.K. and N.S. contributed to critical discussion; S.Y. conceived and designed the experiments, and wrote the paper; M.F. designed the experiments; M.T. conceived and contributed materials and analysis.

Acknowledgments: This work was supported by JSPS KAKENHI Grant Number JP15K13554.

Conflicts of Interest: The authors declare no conflicts of interest.

References

1. Tanaka, T. Collapse of gels and the critical endpoint. *Phys. Rev. Lett.* **1978**, *40*, 820. [CrossRef]
2. Tanaka, T.; Fillmore, D.J. Kinetics of swelling of gels. *J. Chem. Phys.* **1979**, *70*, 1214–1218. [CrossRef]
3. Tanaka, T.; Fillmore, D.; Sun, S.T.; Nishio, I.; Swislow, G.; Shah, A. Phase transitions in ionic gels. *Phys. Rev. Lett.* **1980**, *45*, 1636. [CrossRef]
4. Tanaka, T.; Sun, S.T.; Hirokawa, Y.; Katayama, S.; Kucera, J.; Hirose, Y.; Amiya, T. Mechanical instability of gels at the phase transition. *Nature* **1987**, *325*, 796. [CrossRef]
5. Sato, M.E.; Tanaka, T. Kinetics of discontinuous volume—Phase transition of gels. *J. Chem. Phys.* **1988**, *89*, 1695–1703. [CrossRef]
6. Miki, H.; Yagihara, S.; Mukai, S.A.; Tokita, M. Swelling equilibrium of a gel in binary mixed solvents. *Prog. Colloid Polym. Sci.* **2009**, *136*, 101–106.
7. Hahn, E.L. Spin echoes. *Phys. Rev.* **1950**, *80*, 580. [CrossRef]
8. Stejskal, E.O.; Tanner, J.E. Spin diffusion measurements: Spin echoes in the presence of a time-dependent field gradient. *J. Chem. Phys.* **1965**, *42*, 288–292. [CrossRef]
9. Tokita, M.; Miyoshi, T.; Takegoshi, K.; Hikichi, K. Probe diffusion in gels. *Phys. Rev. E* **1996**, *53*, 1823. [CrossRef]
10. Tokita, M. Transport phenomena in gel. *Gels* **2016**, *2*, 17. [CrossRef]
11. Matsukawa, S.; Ando, I. A study of self-diffusion of molecules in polymer gel by pulsed-gradient spin—Echo 1H NMR. *Macromolecules* **1996**, *29*, 7136–7140. [CrossRef]
12. Matsukawa, S.; Yasunaga, H.; Zhao, C.; Kuroki, S.; Kurosu, H.; Ando, I. Diffusion processes in polymer gels as studied by pulsed field-gradient spin-echo NMR spectroscopy. *Prog. Polym. Sci.* **1999**, *24*, 995–1044. [CrossRef]
13. Matsukawa, S.; Sagae, D.; Mogi, A. Molecular diffusion in polysaccharide gel systems as observed by NMR. *Progr. Colloid Polym. Sci.* **2009**, *136*, 171–176.
14. Yagihara, S.; Miura, N.; Hayashi, Y.; Miyairi, H.; Asano, M.; Yamada, G.; Shinyashiki, N.; Mashimo, S.; Umehara, T.; Tokita, M.; et al. Microwave dielectric study on water structure and physical properties of aqueous systems using time domain reflectometry with flat-end cells. *Subsurf. Sens. Technol. Appl.* **2001**, *2*, 15–30. [CrossRef]
15. Yamada, G.; Hashimoto, T.; Morita, T.; Shinyashiki, N.; Yagihara, S.; Tokita, M. Dielectric study on dynamics for volume phase transition of PAAm gel in acetone-water system. *Trans. Mater. Res. Soc. Jpn.* **2001**, *26*, 701–704.
16. Cole, R.H. Evaluation of dielectric behavior by time domain spectroscopy. I. Dielectric response by real time analysis. *J. Phys. Chem.* **1975**, *79*, 1459–1469. [CrossRef]
17. Cole, R.H. Evaluation of dielectric behavior by time domain spectroscopy. II. Complex permittivity. *J. Phys. Chem.* **1975**, *79*, 1469–1474. [CrossRef]
18. Cole, R.H.; Mashimo, S.; Winsor, P., IV. Evaluation of dielectric behavior by time domain spectroscopy. 3. Precision difference methods. *J. Phys. Chem.* **1980**, *84*, 786–793. [CrossRef]
19. Yang, M.; Zhao, K. Influence of the structure on the collapse of poly (N-isopropylacrylamide)-based microgels: An insight by quantitative dielectric analysis. *Soft Matter* **2016**, *12*, 4093–4102. [CrossRef] [PubMed]
20. Yang, M.; Liu, C.; Zhao, K. Concentration dependent phase behavior and collapse dynamics of PNIPAM microgel by dielectric relaxation. *Phys. Chem. Chem. Phys.* **2017**, *19*, 15433–15443. [CrossRef] [PubMed]
21. Yang, M.; Liu, C.; Lian, Y.; Zhao, K.; Zhu, D.; Zhou, J. Relaxations and phase transitions during the collapse of a dense PNIPAM microgel suspension—Thorough insight using dielectric spectroscopy. *Soft Matter* **2017**, *13*, 2663–2676. [CrossRef] [PubMed]
22. Yagihara, S.; Asano, M.; Kosuge, M.; Tsubotani, S.; Imoto, D.; Shinyashiki, N. Dynamical behavior of unfreezable molecules restricted in a frozen matrix. *J. Non-Cryst. Solids* **2005**, *351*, 2629–2634. [CrossRef]
23. Shinyashiki, N.; Imoto, D.; Yagihara, S. Broadband dielectric study of dynamics of polymer and solvent in poly (vinyl pyrrolidone)/normal alcohol mixtures. *J. Phys. Chem. B* **2007**, *111*, 2181–2187. [CrossRef] [PubMed]
24. Mashimo, S.; Miura, N.; Umehara, T.; Yagihara, S.; Higasi, K. The structure of water and methanol in p-dioxane as determined by microwave dielectric spectroscopy. *J. Chem. Phys.* **1992**, *96*, 6358–6361. [CrossRef]

25. Schrödle, S.; Hefter, G.; Buchner, R. Dielectric spectroscopy of hydrogen bond dynamics and microheterogenity of water + dioxane mixtures. *J. Phys. Chem. B* **2007**, *111*, 5946–5955. [CrossRef] [PubMed]
26. Cole, R.H. Correlation function theory of dielectric relaxation. *J. Chem. Phys.* **1965**, *42*, 637–643. [CrossRef]
27. Davidson, D.W.; Cole, R.H. Dielectric relaxation in glycerol, propylene glycol, and n-propanol. *J. Chem. Phys.* **1951**, *19*, 1484–1490. [CrossRef]
28. Kaatze, U.; Pottel, R.; Schäfer, M. Dielectric spectrum of dimethyl sulfoxide/water mixtures as a function of composition. *J. Phys. Chem.* **1989**, *93*, 5623–5627. [CrossRef]
29. Lu, Z.; Manias, E.; Macdonald, D.D.; Lanagan, M. Dielectric relaxation in dimethyl sulfoxide/water mixtures studied by microwave dielectric relaxation spectroscopy. *J. Phys. Chem. A* **2009**, *113*, 12207–12214. [CrossRef] [PubMed]
30. Havriliak, S.; Negami, S. A complex plane representation of dielectric and mechanical relaxation processes in some polymers. *Polymer* **1967**, *8*, 161–210. [CrossRef]
31. Yagihara, S.; Nozaki, R.; Takeishi, S.; Mashimo, S. Evaluation of dielectric permittivity by dc transient current. *J. Chem. Phys.* **1983**, *79*, 2419–2422. [CrossRef]
32. Nozaki, R.; Mashimo, S. Dielectric relaxation measurements of poly (vinyl acetate) in glassy state in the frequency range 10^{-6}–10^6 Hz. *J. Chem. Phys.* **1987**, *87*, 2271–2277. [CrossRef]
33. Kato, S.; Kita, R.; Shinyashiki, N.; Yagihara, S.; Fukuzaki, M. Diffusion phenomenon of molecules in 1,4-dioxane-water system observed by 1H NMR method. *Proc. Sch. Sci. Tokai Univ.* **2009**, *44*, 53–61.
34. De Gennes, P.G. *Scaling Concepts in Polymer Physics*; Cornell University Press: Ithaca, NY, USA, 1979.
35. Yagihara, S.; Oyama, M.; Inoue, A.; Asano, M.; Sudo, S.; Shinyashiki, N. Dielectric relaxation measurement and analysis of restricted water structure in rice kernels. *Meas. Sci. Technol.* **2007**, *18*, 983. [CrossRef]
36. Kundu, S.K.; Yagihara, S.; Yoshida, M.; Shibayama, M. Microwave dielectric study of an oligomeric electrolyte gelator by time domain reflectometry. *J. Phys. Chem. B* **2009**, *113*, 10112–10116. [CrossRef] [PubMed]
37. Maruyama, Y.; Numamoto, Y.; Saito, H.; Kita, R.; Shinyashiki, N.; Yagihara, S.; Fukuzaki, M. Complementary analyses of fractal and dynamic water structures in protein-water mixtures and cheeses water structures. *Colloids Surf. A Physicochem. Eng. Asp.* **2014**, *440*, 42–48. [CrossRef]
38. Ryabov, Y.E.; Feldman, Y.; Shinyashiki, N.; Yagihara, S. The symmetric broadening of the water relaxation peak in polymer–water mixtures and its relationship to the hydrophilic and hydrophobic properties of polymers. *J. Chem. Phys.* **2002**, *116*, 8610–8615. [CrossRef]
39. Feldman, Y.; Puzenko, A.; Ryabov, Y. Non-Debye dielectric relaxation in complex materials. *Chem. Phys.* **2002**, *284*, 139–168. [CrossRef]
40. Yagihara, S.; Asano, M.; Shinyashiki, N. Broadband dielectric spectroscopy study on hydration of cement and some aqueous solution and dispersion systems. *Conf. Proc. Int. Soc. Electromagn. Aquametry* **2007**, *7*, 11–18.
41. Holz, M.; Heil, S.R.; Sacco, A. Temperature-dependent self-diffusion coefficients of water and six selected molecular liquids for calibration in accurate 1H NMR PFG measurements. *Phys. Chem. Chem. Phys.* **2000**, *2*, 4740–4742. [CrossRef]

gels

MDPI

Article

Emergence of Wrinkles during the Curing of Coatings

Michiko Shimokawa [1,*], Hikaru Yoshida [1], Takumi Komatsu [1], Rena Omachi [2] and Kazue Kudo [2]

[1]   Fukuoka Institute of Technology, Fukuoka 811-0295, Japan;
     s13e2070@bene.fit.ac.jp (H.Y.); s13e2027@bene.fit.ac.jp (T.K.)
[2]   Department of Computer Science, Ochanomizu University, Tokyo 112-8610, Japan;
     omachi.rena@is.ocha.ac.jp (R.O.); kudo@is.ocha.ac.jp (K.K.)
*   Correspondence: shimokawa@fit.ac.jp; Tel.: +81-92-606-5342

Received: 15 February 2018; Accepted: 29 April 2018; Published: 3 May 2018

Abstract: Wrinkles often emerge on a paint layer when a second coat of paint is applied on an already-coated substrate. Wrinkle formation occurs when the first layer absorbs organic solvent from the second layer. We set up experiments to mimic the double-coating process, focusing on the interaction between a paint layer and an organic solvent. In the experiments, we investigated the characteristic wavelengths of the wrinkles and the time of wrinkle emergence. We employed a simple model to explain the wrinkle emergence and performed numerical simulations. The linear stability analysis of the model provides a relation between the wavelengths and the characteristic timescale that agrees reasonably well with our experimental data as well as numerical results. Our results indicate that compression of the layer due to swelling and delamination are both important factors in the formation of wrinkles.

Keywords: paint coating; wrinkle; swelling; buckling

1. Introduction

Double coating, in which paint is applied over an already-coated substrate, is often used to avoid unevenness in the paint layers. In spite of double coating, however, wrinkles sometimes emerge in the drying process, if (1) the elapsed time between the first and second coatings is too short, or (2) the coat thickness is too great [1,2]. The formation of wrinkles has been studied in many fields, such as engineering, material science, chemistry, and physics [3]. The fundamental process of wrinkle formation, however, is not yet fully understood. Below, we focus on case (1) above and investigate the formation of wrinkles from the viewpoint of the mechanical stability of the paint layer.

In a double-coating process, two layers of paint are produced by the first and second coatings. Deformation of the first layer, underlying the second layer, leads to the formation of wrinkles observed at the surface of the second layer. A resin paint, which includes a polymer and an organic solvent, is often used in the painting process. The mechanism of wrinkle formation by a resin paint is thought to be as follows: A polymerization reaction proceeds in the layer after the first application, and the stiffness of the layer increases as it cures [4]. When a second coating is applied, the organic solvent, which is an ingredient in the second coating of paint, penetrates into the first layer. Exposure to the organic solvent causes the polymerized first layer to swell [5]. The first layer is easily swollen when the elapsed time between the first and second coatings is too short, because of incomplete polymerization of the first layer [6]. The swelling induced by absorption of the solvent thus causes deformation of the first layer, and wrinkle formation at the surface of the second layer is due to the resulting deformation of the underlying first layer. Most previous experiments on double coating have focused on the top (second) layer rather than the first layer [1,7], and the effect by the deformation of only the first layer has not been investigated quantitatively.

In this paper, we propose an experiment in which an organic solvent is applied to the surface of the first layer to mimic the double-coating process. We can then observe the deformation of the first layer directly in the experiment, without the complicated effects of the deformation and polymerization of the second layer. The emergence of wrinkles in this experiment is due solely to the deformations caused by the instability of the first layer. We investigate the characteristic length scales, i.e., the wavelengths, of the wrinkles, together with the characteristic timescale that characterizes wrinkle formation. The characteristic lengths and the timescale depend on the elapsed time T between the application of the first coating and the application of the organic solvent. In order to investigate the emergence of wrinkles, we employ a simple model that includes both the effects of buckling due to the swelling of the layer and delamination of the layer from the substrate. The model provides a relation between the wavelengths and the characteristic timescale. The relation is demonstrated as novel types of plots of our experimental and numerical results. The result indicates that the swelling of the layer and its delamination from the substrate cause the instability of the layer that leads to the emergence of wrinkles. In the following sections, we discuss experimental and numerical data in detail and consider the process of wrinkle formation by means of a model.

2. Results and Discussion

2.1. Experimental Results

2.1.1. Buckle Formation

Figure 1 shows the deformations of the paint layer at the following times t after application of the organic solvent: (a) $t = 58$ s, (b) 63 s, (c) 90 s, (d) 120 s, (e) 150 s, and (f) 300 s, all for experiments at the fixed time $T = 24$ h after the application of the paint layer. The drop of organic solvent spreads into a circular shape approximately 11 mm in diameter. Short-scale wrinkles first emerge at $t = 58$ s (Figure 1a). Shortly after that, at $t = 63$ s, larger-scale wrinkles appear (Figure 1b), and small bumps appear randomly at $t = 90$ s (Figure 1c). The bump amplitudes are much larger than those of wrinkles observed at earlier times. The amplitude of the bumps increase with t, and delamination of the layer from the substrate occurs. In the process, some bumps coalesce with other bumps (Figure 1d,e). The coalescence repeats, and buckles emerge, as shown in Figure 1f. The pattern of buckles does not change after $t = 300$ s.

We focus here on the deformations of the paint layer that occur in experiments for several different values of the curing time T. Figure 2 shows snapshots for (a) $T = 1$ h, (b) 24 h, (c) 56 h, and (d) 64 h. These images were all obtained at $t = 10$ min. Buckles emerge only in experiments for $T = 24$ h (Figure 2b). The paint layer is melted by the organic solvent in experiments for $T = 1$ h (Figure 2a). Several bumps appear in experiments for $T = 56$ h (Figure 2c), but they vanish at $t = 30$ min, and a layer with a smooth surface remains. In experiments for $T = 64$ h, the surface of the layer remains smooth and does not change with time (Figure 2d). The experiments for $T = 56$ and 64 h both result in smooth surfaces, even after the application of the organic solvent, but the processes by which the smooth surfaces are produced are different. These results show that buckles emerge only for a limited range of the curing times T. This behavior is similar to the results obtained in previous experiments with double coatings [1,7].

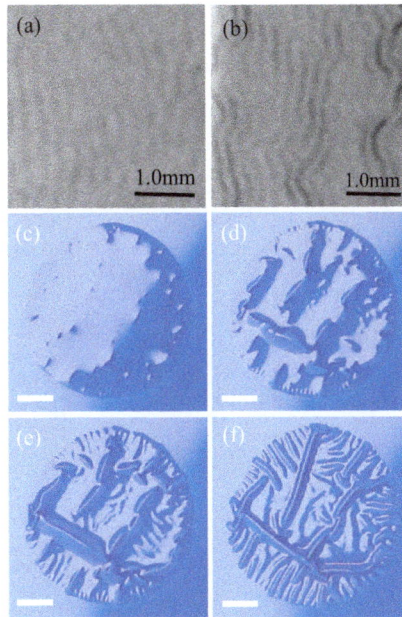

Figure 1. Deformations of the paint layer at (**a**) $t = 58$ s, (**b**) 63 s, (**c**) 90 s, (**d**) 120 s, (**e**) 150 s, and (**f**) 300 s, where $t = 0$ is the time when the organic solvent is applied to the layer. The contrast of (**a,b**) is modified only for clearer demonstration of wrinkles. These photographs all apply to experiments for which the curing time $T = 24$ h. The solid lines in panels (**a,b**) are 1.0-mm-scale bars, and those in panels (**c–f**) are 3.0-mm-scale bars.

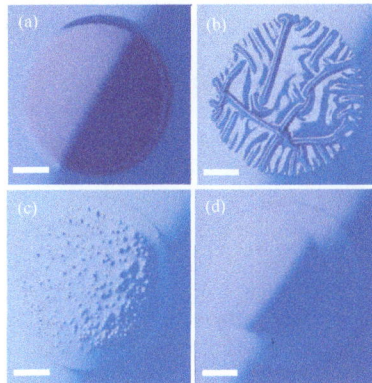

Figure 2. Deformations of layers obtained in experiments for (**a**) $T = 1$ h, (**b**) 24 h, (**c**) 56 h, and (**d**) 64 h, where T is the time elapsed between the application of the paint layer and the application of the drop of organic solvent. These images were taken at $t = 600$ s after the application of the organic solvent. Solid lines in the photos are 3.0-mm-scale bars.

2.1.2. Characteristic Spatial Scales and Timescales for the Formation of Wrinkles

Figure 3a shows the short-scale wavelength λ_s and the wavelength λ_w of larger-scale wrinkles obtained from experiments with $T = 24$ h, which were mentioned in Section 2.1.1. Small structures with wavelengths λ_s appear first, and wrinkles with wavelengths λ_w appear subsequently. The quantity λ_w is the maximum wavelength observed before the wrinkles coalesce. Figure 3b,c shows the values of λ_s and λ_w, respectively, in experiments for several different values of T. We assume that they can be fitted by the linear functions

$$\lambda_s = c_{s1}T + c_{s2}, \tag{1}$$
$$\lambda_w = c_{w1}T + c_{w2} \tag{2}$$

with the fitting parameters $c_{s1} = 0.41 \times 10^{-2}$, $c_{s2} = 0.17$, $c_{w1} = 0.44 \times 10^{-1}$, and $c_{w2} = 0.37$. Using Equations (1) and (2), we can determine λ_s and λ_w for any value of T.

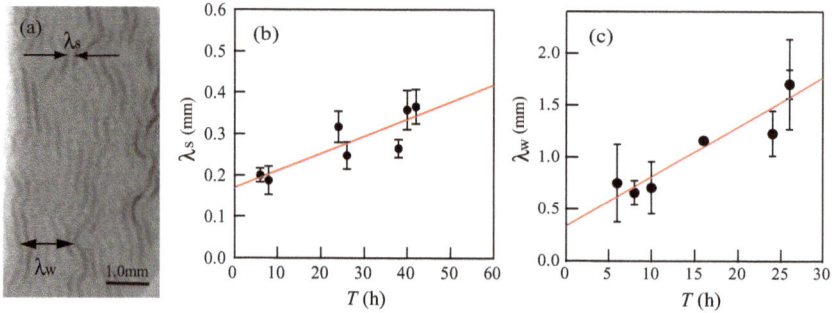

Figure 3. (a) Snapshot taken at $t = 63$ s in the experiment with elapsed time $T = 24$ h. The short-scale wavelength λ_s of the small wrinkles and the wavelength λ_w of the larger-scale wrinkles are indicated. The solid line in the photo shows a 1.0-mm-scale bar. Panels (b,c) show the quantities λ_s and λ_w obtained from our experiments for several different values of T. The closed circles are the experimental data, and the solid lines in (b,c) are the fitted lines given by Equations (1) and (2), respectively.

Next, we investigate the characteristic timescale τ_{ex}, which turns out to be inversely related to the growth rate of the wrinkles. We define τ_{ex} as the time elapsed between the application of the organic solvent and the appearance of bumps of 0.2 mm in diameter. As shown in Figure 4, τ_{ex} increases with T. The timescale τ_{ex} is larger than the times at which λ_s and λ_w are measured. After the initial growth of patterns with wavelengths λ_s and λ_w, the coalescence of wrinkles is caused by nonlinear effects. Coalescence leads to a change in the characteristic length of pattern deformation. The time when such a change occurs is proportional to the timescale τ_{ex} [8].

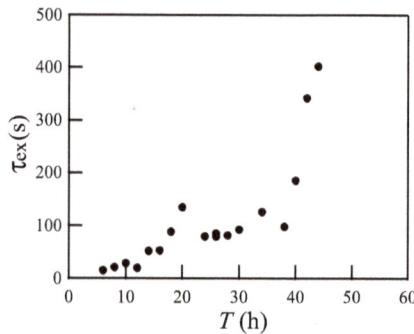

Figure 4. Relationship between T and τ_{ex}, where τ_{ex} is the time elapsed between the application of the organic solvent and the appearance of bumps of 0.2 mm diameter.

2.2. Model

We next introduce a simple model for the buckling that is observed in our experiments. Consider the coating of paint to be an elastic thin film that is attached adhesively to a solid substrate (Figure 5). Because of the absorption of the organic solvent, the elastic film swells, producing compression stress. Suppose that the elastic film exists on a flat substrate, whose surface corresponds to the x_1-x_2 plane, and let the x_3 axis be normal to the surface of the substrate. The total energy F_{tot} of this system consists of the elastic strain energy in the film and the interfacial traction energy between the film and substrate:

$$F_{tot} = \iint (f_{film} + f_{int})\, dx_1 dx_2, \tag{3}$$

where f_{film} and f_{int} are the energies per unit area of the film and the interface, respectively.

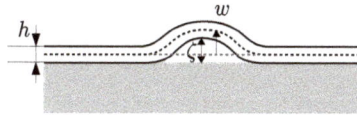

Figure 5. Schematic of a thin film on a substrate. The thickness of film is h. The mid-plane displacement and the distance between the substrate and film are denoted by w and ζ, respectively.

According to the Föppl–von Kármán plate theory, the elastic energy per unit area in a film of thickness h is given by [8–12]

$$f_{film} = \int_{-h/2}^{h/2} \frac{1}{2} \sigma_{\alpha\beta}^{el} \varepsilon_{\alpha\beta}^{el}\, dx_3, \tag{4}$$

$$\sigma_{\alpha\beta}^{el} = \frac{2\mu}{1-\nu}\left[(1-\nu)\varepsilon_{\alpha\beta}^{el} + \nu\varepsilon_{\gamma\gamma}^{el}\delta_{\alpha\beta}\right], \tag{5}$$

$$\varepsilon_{\alpha\beta}^{el} = e_{\alpha\beta} - x_3 \frac{\partial^2 w}{\partial x_\alpha \partial x_\beta}. \tag{6}$$

Greek subscripts refer to in-plane coordinates x_1 or x_2, and repeated Greek subscripts indicate summation over indices 1 and 2. The parameters μ and ν are the shear modulus and Poisson ratio of the film, respectively. Mid-plane displacements in the in-plane and x_3 directions are denoted by u_α and w, respectively. Supposing the film to be under equibiaxial stress, we take the initial in-plane strain to be $\varepsilon_0\delta_{\alpha\beta}$. Then,

$$e_{\alpha\beta} = \frac{1}{2}\left(\frac{\partial u_\alpha}{\partial x_\beta} + \frac{\partial u_\beta}{\partial x_\alpha}\right) + \frac{1}{2}\frac{\partial w}{\partial x_\alpha}\frac{\partial w}{\partial x_\beta} - \varepsilon_0\delta_{\alpha\beta}. \tag{7}$$

To express the interfacial traction energy between the film and substrate, we use the cohesive zone model [13–15]. The interfacial energy per unit area is then

$$f_{int} = \int_0^\zeta T_n(z)\, dz, \tag{8}$$

where ζ is the distance between the substrate and film, and T_n is the normal traction. When the film thickness is constant, $\zeta = w$. We represent the normal traction as

$$T_n(\zeta) = \Gamma_n\zeta \exp\left(-\frac{\zeta}{\delta_n}\right), \tag{9}$$

where $\Gamma_n \equiv \gamma_n/\delta_n^2$. The parameters γ_n and δ_n are the normal interfacial toughness and the characteristic length of a normal displacement jump, respectively.

The total energy F_{tot} is thus expressed in terms of the displacements u_α and w. Equilibrium states must satisfy $\delta F_{\text{tot}}/\delta w = 0$ and $\delta F_{\text{tot}}/\delta u_\alpha = 0$. However, instead of solving $\delta F_{\text{tot}}/\delta w = 0$, we employ the time-dependent Ginzburg–Landau equation, which is often used in dynamical systems,

$$\frac{\partial w}{\partial t} = -\eta \frac{\delta F_{\text{tot}}}{\delta w}, \tag{10}$$

where η is a constant related to the characteristic relaxation time. Scaling all lengths by h, times by $h/(\mu\eta)$, the nondimensional equation and γ_n by $h\mu$ in Equation (10), we obtain

$$\frac{\partial w}{\partial t} = -\frac{1}{6(1-\nu)} \nabla^2 \nabla^2 w + \frac{\partial N_\beta}{\partial x_\beta} - T'_n, \tag{11}$$

where the variables are dimensionless, T'_n is the nondimensional form of Equation (9), and

$$N_\beta = \sigma_{\alpha\beta} \frac{\partial w}{\partial x_\alpha}, \tag{12}$$

$$\sigma_{\alpha\beta} = \frac{2}{1-\nu} \left[(1-\nu)e_{\alpha\beta} + \nu e_{\gamma\gamma}\delta_{\alpha\beta} \right]. \tag{13}$$

The in-plane displacements u_α included in $e_{\alpha\beta}$ are obtained from the equation $\delta F_{\text{tot}}/\delta u_\alpha = 0$.

2.3. Linear Stability Analysis

A linear stability analysis of Equation (11) provides some insight into the condition of buckling. Linearizing Equation (11) around $w = 0$, and taking the Fourier transform of the linearized equation, we obtain

$$\frac{\partial \tilde{w}(k)}{\partial t} = g(k)\tilde{w}(k), \tag{14}$$

where $\tilde{w}$ is the Fourier transform of w, and k is the wavenumber. The linear growth rate g is given by

$$g(k) = -\frac{1}{6(1-\nu)}[k^2 - 6(1+\nu)\varepsilon_0]^2 + \frac{6(1+\nu)^2}{1-\nu}\varepsilon_0^2 - \Gamma_n. \tag{15}$$

Unstable modes, which cause deformations in the layer, appear when $g(k) > 0$; in other words,

$$\Gamma_n < \frac{6(1+\nu)^2}{1-\nu}\varepsilon_0^2. \tag{16}$$

This equation shows that wrinkles emerge above a certain threshold of stress. The existence of the threshold is consistent with the experimental results shown in Figure 2, which indicate that buckles emerge under an upper limit of T, since Γ_n and ε_0 depend on T. Equation (15) shows that the wavenumber of the fastest-growing mode is

$$k_f = \sqrt{6(1+\nu)\varepsilon_0}. \tag{17}$$

The growth rate of the fastest-growing mode is inversely proportional to the timescale,

$$\tau_f = \left[\frac{6(1+\nu)^2}{1-\nu}\varepsilon_0^2 - \Gamma_n \right]^{-1}. \tag{18}$$

2.4. Numerical Simulations

Numerical simulation is useful for demonstrating that our simple model does reproduce buckling of the film. Simulated patterns of the displacement w are shown in Figure 6. In the initial states, we take $w = 0$ plus a small amount of noise, and we impose periodic boundary conditions on a 256×256 grid system. The length of a side corresponds to about 6.8 mm for $h = 0.13$ mm. The values of the side length and h are close to experimental ones. In Figure 6, we take $\varepsilon_0 = 1.2$, $\Gamma_n = 1.4$. The other parameters used in the following simulations are $\nu = 0.3$ and $\delta_n = 0.5$.

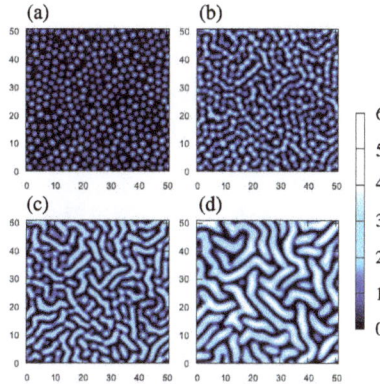

Figure 6. Snapshots of numerical simulations at (**a**) $t = 120$, (**b**) 160, (**c**) 200, and (**d**) 400. The color scale illustrates the mid-plane displacement w. Panels (**a,b**) correspond to (**c–f**) of Figure 1, respectively. The length of a side of a snapshot corresponds to 6.8 mm when the film thickness is $h = 0.13$ mm.

Some characteristics of the snapshots in Figure 6 look similar to those of the experiments in Figure 1c–f. Small bumps appear at an early stage (Figure 6a). The amplitudes of the bumps grow with time, and some bumps coalesce with others (Figure 6b,c). However, the amplitudes continue to grow in the simulations (Figure 6d), which is significantly different from the experiments. This indicates that our model is not yet adequate to explain the nonlinear effects in the actual experiments.

The phase diagram shown in Figure 7 illustrates the numerical verification of the wrinkle-emerging condition. The linear stability analysis suggests that wrinkles should appear above the solid curve, which is given by Equation (16). Numerical data shown as symbols demonstrate that the analysis is sufficiently valid.

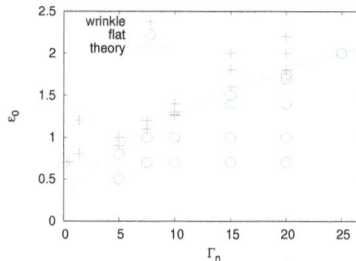

Figure 7. Phase diagram of wrinkle emergence. The horizontal and vertical axes are the parameter relating to the interfacial toughness and the initial in-plane strain, respectively. The solid curve corresponds to Equation (16), which is given by the linear stability analysis. Symbols are data of numerical simulations.

Figure 8a shows the time evolution of the root-mean-square (RMS) of w,

$$W_{\text{RMS}} = \sqrt{\frac{\int w(\mathbf{r})^2 d^2 r}{L^2}}, \tag{19}$$

where L is the system size. The parameter values except for ε_0 are taken as the same as Figure 6. This is also interpreted as the time evolution of the average amplitude A of wrinkles, since $W_{\text{RMS}} = \sqrt{1/2}A$, where we suppose a stripe form of $w = A \sin(k_f x_1)$. The arrows in Figure 8a indicate the initial-growth time τ which is the transition time from the initial-growth regime to the coarsening one. Let us here estimate the amplitude at the initial-growth time. In the coarsening regime, we can assume that the interfacial energy becomes negligible and that the time evolution slows down significantly. Thus, setting $\partial w / \partial t = T'_n = 0$ in Equation (11), we have

$$\sigma_{11} = -\frac{k_f^2}{6(1-\nu)}. \tag{20}$$

On the other hand, calculating the spacial average of σ_{11} given by Equation (13) leads to

$$\bar{\sigma}_{11} = \frac{2}{1-\nu} \left[\frac{A^2}{4} k_f^2 - (1+\nu)\varepsilon_0 \right] = \frac{(3A^2 - 2)k_f^2}{6(1-\nu)}. \tag{21}$$

Equating Equtaions (20) and (21), we have $A = \sqrt{1/3}$. The initial-growth time τ is determined as the time when $A = \sqrt{1/3}$, and thus, $W_{\text{RMS}} = 1/\sqrt{6}$.

The average wavelength of wrinkles is $\lambda = 2\pi/\bar{k}$, where

$$\bar{k} = \sqrt{\frac{\int k^2 |\tilde{w}(\mathbf{k})|^2 d^2 k}{\int |\tilde{w}(\mathbf{k})|^2 d^2 k}}. \tag{22}$$

Here, $\tilde{w}$ is the Fourier transform of $w - \bar{w}$, where $\bar{w}$ is the spacial average of w. The time evolution of the average wavelength is shown in Figure 8b. In the initial-growth regime, the average wavelength is approximately equal to (or slightly larger than) that of the fastest-growing mode, which is indicated by the arrows in Figure 8b. The wavelength λ_f of the fastest-growing mode is estimated from the linear stability analysis and evaluated as $\lambda_f = 2\pi/k_f$, where k_f is given by Equation (17).

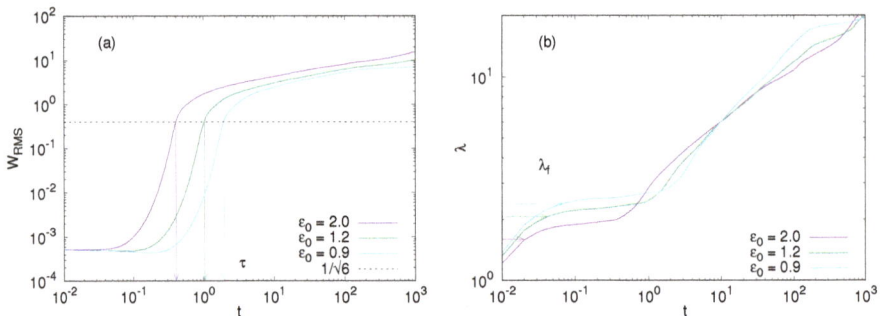

Figure 8. Time evolution of (a) the root mean square of w (related to the amplitude of wrinkles,b) the average wavelength of wrinkles. The arrows in (a) indicate the time when $W_{\text{RMS}} = 1/\sqrt{6}$. The arrows in (b) indicate the wavelength $\lambda_f = 2\pi/k_f$ of the fastest mode, which is given by Equation (17).

2.5. Correspondence between Experimental and Numerical Results

We here rewrite the time scale τ_f in other forms to examine experimental and numerical results by means of the linear stability analysis. Suppose that the growth rate of a certain unstable mode k_1 is $g(k_1) = C$, where C is a positive constant. Using Equations (15), (17) and (18), we have

$$k_f^{\,2} - k_1^2 = \sqrt{6(1-\nu)(\tau_f^{-1} - C)}, \tag{23}$$

where $k_1 < k_f$. Equation (23) leads to

$$\tau_f = \frac{6(1-\nu)}{(k_f^2 - k_1^2)^2 + C'} \propto \left[\left(\frac{1}{\lambda_f^2} - \frac{1}{\lambda_1^2} \right)^2 + C' \right]^{-1}, \tag{24}$$

where $k_f = 2\pi/\lambda_f$, $k_1 = 2\pi/\lambda_1$, and C' is a constant. For the minimum wavenumber $k_{\min}$ with a non-negative growing rate, $g(k_{\min}) = 0$, and thus, $C = C' = 0$. Then, Equtaions (23) and (24) turn to be

$$\tau_f = \frac{6(1-\nu)}{(k_f^2 - k_{\min}^2)^2} \propto (k_f^2 - k_{\min}^2)^{-2}, \tag{25}$$

where

$$k_f^2 - k_{\min}^2 = \sqrt{6(1-\nu)\left(\frac{6(1+\nu)^2}{1-\nu}\varepsilon_0^2 - \Gamma_n \right)}. \tag{26}$$

Equations (24) and (25) are useful to examine experimental and numerical results, respectively.

We first examine experimental results, using Equation (24). We assume that λ_f and λ_1 in Equation (24) correspond to λ_s and λ_w in Figure 3, respectively. This assumption implies that structures with wavelengths λ_s and λ_w appear in the linear-instability region and that λ_s and λ_w correspond to unstable modes of pattern formation. We also assume that $\tau_{ex} \propto \tau_f$ [8].

The closed circles in Figure 9 show experimental data about the relation between τ_{ex} and $(1/\lambda_s^2) - (1/\lambda_w^2)$. Values of λ_s and λ_w are obtained using Equations (1) and (2). The experimental data agree reasonably closely with the line given by Equation (24), which is shown as a solid line. The exponent of the fitted line is -2, and C' of Equation (24) nearly vanishes for the fitted line.

Figure 10 shows the numerical counterparts. The initial-growth time τ is plotted as a function of $k_f^2 - k_{\min}^2$. The initial-growth time is defined as the time when W_{RMS} reaches $1/\sqrt{2}$ (See Figure 8), and its value is obtained from numerical simulations with different combinations of ε_0 and Γ_n. The parameters used in the simulation gives $k_f^2 - k_{\min}^2$ through Equation (26). Assuming that the initial-growth time τ is proportional to τ_f of Equation (25), we fit the line given by Equation (25) to the numerical data. The fitted line reasonably agrees with the numerical data.

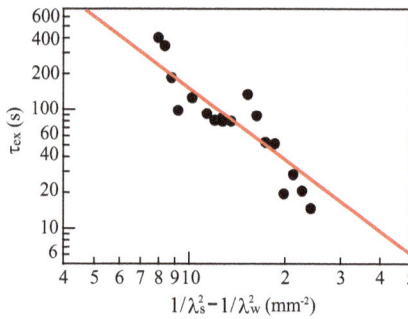

Figure 9. The relationship between τ_{ex} and $(1/\lambda_s^2) - (1/\lambda_w^2)$, where τ_{ex} is the time between the application of the organic solvent and the appearance of bumps of 0.2 mm diameter. The quantities λ_s and λ_w are the wavelengths of the small wrinkles that first appear and the maximum wavelength observed before the wrinkles coalesce, respectively. The closed circles are the experimental results, and the solid line is the fit from Equation (24).

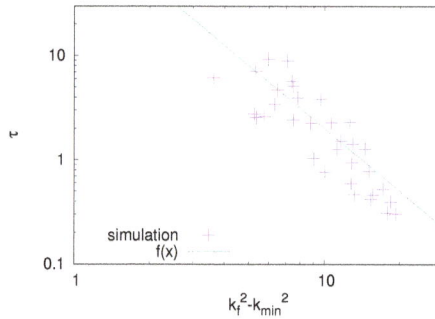

Figure 10. Initial-growth time τ as a function of $k_f^2 - k_{min}^2$, where k_f and k_{min} denote the wavenumber of the fastest-growing mode and the minimum wavenumber with a non-negative growth rate, respectively. Symbols are the numerical results, and the solid line is the fit from Equation (25).

3. Conclusions

The objective of this paper has been to understand the emergence of wrinkles at the surface of a coating following the application of an organic solvent. The instability at the surface of the layer leads to the emergence of wrinkles. We investigated the characteristic lengths of the wrinkles and the characteristic timescale for wrinkle emergence in experiments and numerical simulations. The linear stability analysis of our simple model supports the experimental and numerical results. Although the simple model suitably explains the emergence of wrinkles, we will need a more realistic model to investigate the coarsening of wrinkles and time evolution of wrinkle patterns. For example, the film thickness, the strain induced by volume expansion and the interfacial traction vary with time in experiments as the solvent evaporates. Those effects should be included in the model to investigate wrinkle patterns beyond the linear-stability regime.

Our results indicate that the initial strain ε_0 and the interfacial toughness Γ_n depend on the curing time T. Although the dependencies have not been specified yet, our results will be useful especially in engineering. For example, even if T is unknown, we can estimate ε_0 and Γ_n from Equtaions (17) and (18) by measuring the wavelength of wrinkles and τ_{ex}. Those parameters are essential for the control of wrinkle formation.

Gels 2018, 4, 41

We conclude that (1) buckling due to volume expansion of the layer and (2) delamination of the layer from the substrate are both important for the formation of wrinkles. This conclusion is supported by the linear stability analysis which states that the emergence of wrinkles depends on both the initial strain caused by volume expansion and the normal traction. Experimental results as well as numerical ones show reasonably good agreement with the linear stability analysis.

4. Materials and Methods

4.1. Experimental Method

We used a copper board of 5.0 cm × 5.0 cm square and 1.0 mm thick as the substrate for painting. To control the thickness of the paint layer, we placed two metallic boards facing each other on opposite sides of the copper board, as shown in Figure 11a. The metallic boards are of equal thickness and are slightly thicker than the copper board. We applied a phthalic resin paint (Rubicon1000, No. 837, ISHIKAWA PAINT) to the copper board using a syringe (SS-20ESZ, TERMO), and we spread the paint across the copper board using a metallic bar, producing a layer of relatively uniform thickness. We measured the thickness h of the paint layer using a laser displacement meter (LT9010M, KEYENCE). As shown in Figure 11b, we found that the layer had a nearly uniform thickness with an average value $h = 130 \pm 6$ μm. The coated board was then placed in a constant-temperature oven (NEXAS OFX-70, ASONE) at 30 °C for a time T. After the time T, we applied a 0.02 cm^3 drop of xylene, which is the organic solvent in the paint, on the coated layer. The surface of the layer was photographed with a digital camera (Canon EOS Kiss X4, EF-S 18-55IS) 10 min after the xylene application. It was easy to observe the deformations of the paint layer, since xylene is clear and colorless.

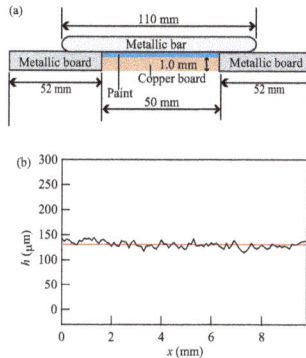

Figure 11. (a) Schematic drawing of the experimental setup for the application of a paint layer. Only the coated copper board is kept in a constant-temperature oven for several hours after the first coating. (b) The surface height h measured after painting.

4.2. Numerical Procedure

We employed a spectral method for numerical simulations. The Fourier transform of Equation (11) is

$$\frac{\partial \tilde{w}}{\partial t} = -Dk^4\tilde{w} - ik_\beta \tilde{N}_\beta - \tilde{T}_n, \tag{27}$$

where $D = 1/[6(1-\nu)]$ and $\tilde{N}_\beta$ and $\tilde{T}_n$ are the Fourier transforms of Equation (12) and of the normal traction T'_n, respectively.

The nonlinear term N_β includes derivatives of u_α. By using the condition $\delta F_{\text{tot}}/\delta u_\alpha = 0$, the Fourier transform of u_α can be written as

$$\tilde{u}_\alpha = \tilde{G}_{\alpha\beta}\tilde{\rho}_\beta, \tag{28}$$

where

$$\tilde{G}_{\alpha\beta} = \frac{1}{1-\nu}\left(\frac{\delta_{\alpha\beta}}{k^2} - \frac{1+\nu}{2}\frac{k_\alpha k_\beta}{k^4}\right), \tag{29}$$

$$\tilde{\rho}_\alpha = \iint\left[(1+\nu)\frac{\partial w}{\partial x_\gamma}\frac{\partial^2 w}{\partial x_\alpha \partial x_\gamma} + (1-\nu)\frac{\partial w}{\partial x_\alpha}\nabla^2 w\right]e^{i\mathbf{k}\cdot\mathbf{r}}dx_1 dx_2. \tag{30}$$

Using Equtaions (28)–(30), we can rewrite Equation (7) in the following form [11,12],

$$e_{\alpha\beta} = \frac{1}{2}\int_{k\neq 0}\left[-i(k_\beta\tilde{G}_{\alpha\gamma} + k_\alpha\tilde{G}_{\beta\gamma})\tilde{\rho}_\gamma\right]\frac{e^{-i\mathbf{k}\cdot\mathbf{r}}}{(2\pi)^2}d^2k + \frac{1}{2}\frac{\partial w}{\partial x_\alpha}\frac{\partial w}{\partial x_\beta} - \varepsilon_0\delta_{\alpha\beta}. \tag{31}$$

In numerical simulations, we used the modified normal traction,

$$T_{\text{n}}(\zeta) = \begin{cases} \Gamma_{\text{n}}\zeta\exp(-\zeta/\delta_{\text{n}}) & \zeta \geq 0, \\ \Gamma'_{\text{n}}\zeta\exp(-\zeta/\delta_{\text{n}}) & \zeta < 0, \end{cases} \tag{32}$$

where Γ'_{n} is a parameter that is sufficiently larger than Γ_{n}. In the simulations, we set $\Gamma'_{\text{n}} = 100\ \Gamma_{\text{n}}$. Although Equation (9) is convenient for linear stability analysis, it is inconvenient for numerical simulations; if Equation (9) was used as the normal traction, areas with $w < 0$ would appear. Since the substrate is solid, negative values of w are not allowed in realistic situations. The modified traction given by Equation (32) enables the calculations to avoid such unrealistic solutions.

For the time evolution, we employed a semi-implicit algorithm: we used first-order backward and forward finite-difference schemes for the linear and nonlinear parts of Equation (27), respectively. The $(n+1)$-th step in the calculation of $\tilde{w}$ is given by

$$\tilde{w}^{(n+1)} = \frac{\tilde{w}^{(n)} - (ik_\beta\tilde{N}_\beta^{(n)} + \tilde{T}_{\text{n}}^{(n)})\Delta t}{1 + Dk^4\Delta t}, \tag{33}$$

where Δt is the time increment.

Author Contributions: K.K. and R.O. performed numerical simulations; K.K. and M.S. designed the analysis; M.S., H.Y. and T.K. performed the experiments; M.S. and K.K. wrote the paper.

Acknowledgments: We would like to thank M. Tokita, S. Ohta, T. Yamaguchi, R. Ushijima for fruitful discussions and suggestions. We also would like to thank Co. Ishikawa Paint in Osaka for demonstration of interesting phenomena observed in coating process. This work was supported by JSPS KAKENHI Grant No. 15K04760.

Conflicts of Interest: The authors declare no conflict of interest.

References

1. Basu, S.K.; Scriven, L.E.; Francis, L.F.; McCormick, A.V. Mechanism of wrinkle formation in curing coatings. *Prog. Org. Coat.* **2005**, *53*, 1–16. [CrossRef]
2. Basu, S.K.; Scriven, L.E.; Francis, L.F.; McCormick, A.V.; Reichert, V.R. Wrinkling of epoxy powder coatings. *J. Appl. Polym. Sci.* **2005**, *98*, 116–129. [CrossRef]
3. Freund, L.B.; Suresh, S. *Thin Film Materials: Stress, Defect Formation and Surface Evolution*; Cambridge University Press: Cambridge, UK, 2004.
4. Takiyama, E. *Handbook of Polyester Resin*; The Nikkan Kogyo Shimbun: Tokyo, Japan, 1988. (In Japanese)
5. Burrell, H. High Polymer Theory of the Wrinkle Phenomenon. *Ind. Eng. Chem.* **1954**, *46*, 2233–2237. [CrossRef]

6. Tanaka, T.; Sun, S.T.; Hirokawa, Y.; Katayama, S.; Kucera, J.; Hirose, Y.; Amiya, T. Mechanical instability of gels at the phase transition. *Nature* **1987**, *325*, 796–798. [CrossRef]

7. Basu, S.K.; Bergstreser, A.M.; Francis, L.F.; Scriven, L.E.; McCormick, A.V. Wrinkling of a two-layer polymeric coating. *J. Appl. Phys.* **2005**, *98*, 063507. [CrossRef]

8. Huang, R.; Im, S.H. Dynamics of wrinkle growth and coarsening in stressed thin films. *Phys. Rev. E* **2006**, *74*, 026214. [CrossRef] [PubMed]

9. Faou, J.Y.; Parry, G.; Grachev, S.; Barthel, E. How Does Adhesion Induce the Formation of Telephone Cord Buckles? *Phys. Rev. Lett.* **2012**, *108*, 116102. [CrossRef] [PubMed]

10. Pan, K.; Ni, Y.; He, L. Effects of interface sliding on the formation of telephone cord buckles. *Phys. Rev. E* **2013**, *88*, 062405. [CrossRef] [PubMed]

11. Ni, Y.; He, L.; Liu, Q. Modeling kinetics of diffusion-controlled surface wrinkles. *Phys. Rev. E* **2011**, *84*, 051604. [CrossRef] [PubMed]

12. Pan, K.; Ni, Y.; He, L.; Huang, R. Nonlinear analysis of compressed elastic thin films on elastic substrates: From wrinkling to buckle-delamination. *Int. J. Solids Struct.* **2014**, *51*, 3715–3726. [CrossRef]

13. Cerda, E.; Mahadevan, L. Geometry and Physics of Wrinkling. *Phys. Rev. Lett.* **2003**, *90*, 074302. [CrossRef] [PubMed]

14. Barenblatt, G.I. The Mathematical Theory of Equilibrium Cracks in Brittle Fracture. *Adv. Appl. Mech.* **1962**, *7*, 55–129.

15. Park, K.; Paulino, G.H. Cohesive Zone Models: A Critical Review of Traction-Separation Relationships Across Fracture Surfaces. *Appl. Mech. Rev.* **2013**, *64*, 060802. [CrossRef]

gels

MDPI

Article

Propagation of Fatigue Cracks in Friction of Brittle Hydrogels

Tetsuo Yamaguchi [1,2,*], Ryuichiro Sato [1] and Yoshinori Sawae [1,2]

[1]   Department of Mechanical Engineering, Kyushu University, 744 Motooka, Nishi-ku,
       Fukuoka 819-0395, Japan; ryu1ro124@gmail.com (R.S.); sawa@mech.kyushu-u.ac.jp (Y.S.)
[2]   International Institute for Carbon-Neutral Energy Research, Kyushu University, 744 Motooka, Nishi-ku,
       Fukuoka 819-0395, Japan
*   Correspondence: yamaguchi@mech.kyushu-u.ac.jp; Tel.: +81-92-802-3072

Received: 14 May 2018; Accepted: 6 June 2018; Published: 8 June 2018

Abstract: In order to understand fatigue crack propagation behavior in the friction of brittle hydrogels, we conducted reciprocating friction experiments between a hemi-cylindrical indenter and an agarose hydrogel block. We found that the fatigue life is greatly affected by the applied normal load as well as adhesion strength at the bottom of the gel–substrate interface. On the basis of in situ visualizations of the contact areas and observations of the fracture surfaces after the friction experiments, we suggest that the mechanical condition altered by the delamination of the hydrogel from the bottom substrate plays an essential role in determining the fatigue life of the hydrogel.

Keywords: hydrogel; friction; fatigue; wear; fracture; crack; adhesion; delamination

1. Introduction

Recently, hydrogels have attracted much attention of both scientists and engineers because of their unique characteristics: they have a low elastic modulus with large deformability and often exhibit extremely low surface friction [1–6]. Because they are similar to natural articular cartilages in structure and properties [7], hydrogels are expected as a candidate material for artificial articular cartilages that could overcome the drawbacks in the present hard-material-based artificial cartilages and reproduce superior characteristics of natural cartilages [8–18]. However, there is a serious problem that we have to resolve; the low fatigue strength of hydrogels against repetitive loadings. To tackle this problem, two different approaches can be considered: material science and mechanics approaches. The former approach is to make trials to synthesize tough hydrogels. In fact, new types of hydrogels with improved toughness have successfully been developed by material scientists in the last decade [19–21]. On the other hand, the latter would be to understand the basic mechanisms behind crack propagation [22–28] and to create novel types of hydrogels adaptive to the mechanical conditions. However, there have been few studies focusing on the fatigue behavior of hydrogels during sliding [29]; the underlying mechanisms and design principles are not well understood at present.

In this paper, we report our fundamental studies on propagation behavior of fatigue cracks of hydrogels in reciprocating friction experiments. In order to facilitate observations, we used agarose hydrogels as a typical example of brittle hydrogels. By performing in situ visualization of frictional contact and observations of fracture surfaces after the friction experiments, we investigated the mechanisms responsible for the propagation of fatigue cracks.

2. Results and Discussion

2.1. Stress-Relaxation Behavior

First of all, we discuss the stress-relaxation behavior of agarose hydrogels on the basis of the results of the unconfined compression tests. As we mention in the experimental section, hydrogels have multiple relaxation processes because of viscoelasticity and water diffusion. In viscoelastic relaxation, the characteristic time is determined by microscopic or mesoscopic processes and is independent of the size of the hydrogel sample. On the other hand, the relaxation time due to diffusive transport should depend on its size. In order to examine the size dependence, we plotted two relaxation times τ_1 and τ_2 against the "system size" $(1/L^2 + 1/W^2)^{-1}$ in Figure 1 (the physical meaning of this term is discussed in Appendix A). It is clearly seen that τ_2 was much larger than τ_1 as well as that τ_2 depended on the system size while τ_1 did not. From these features, we identified τ_1 and τ_2 as the viscoelastic and diffusive relaxation times, respectively.

We then estimated the permeability of the polymer network from the relaxation time τ_2. According to the stress–diffusion (diffusio-mechanical) coupling model [30] (which is equivalent with biphasic lubrication theory [31–33] and the modified version of Tanaka–Fillmore theory [34]), the relaxation time is described by the following equation (see further details in Appendix A):

$$\tau_2 = \frac{1}{D_c} \frac{1}{\pi^2(\frac{1}{L^2} + \frac{1}{W^2})}, \tag{1}$$

where $D_c = \kappa_0(1 - \phi)(K + 4/3G)$ is the collective diffusion constant; K and G are the osmotic and shear moduli, respectively; κ is the permeability; and ϕ is the volume fraction of polymers.

K and G were determined in the following manner. Just after we applied compression (the stress reached the maximum), the gel was regarded as an incompressible material because there was no time for water to come out of the gel. If we neglect the viscoelastic contribution to the stress (with intensity E_1), we obtain the following equation:

$$G = \frac{\sigma_z(t = 0)}{3\epsilon_z} = \frac{E_\infty + E_2}{3}, \tag{2}$$

where $\sigma_z(t)$ and ϵ_z are the the the compressive stress and strain respectively; E_∞ is the relaxed modulus and E_2 is the intensity of the diffusive mode, as introduced in the experimental section. E_∞ and E_2 are estimated from the fitting of the stress-relaxation curves. On the other hand, the normal stress at infinite time is described by

$$\sigma_z(t = \infty) = E_\infty \epsilon_z = \frac{3GK}{K + G/3}\epsilon_z. \tag{3}$$

From Equation (3), we obtain the following relation:

$$K = \frac{GE_\infty}{9G - 3E_\infty}. \tag{4}$$

We approximate $(1 - \phi) \approx 1$ and obtain the expression for the permeability:

$$\kappa = \frac{D_c}{K + 4/3G}. \tag{5}$$

We calculated D_c, K, G, and κ for each sample and then averaged over all the samples. The values estimated were $D_c = 1.5 \times 10^{-7} \pm 9.7 \times 10^{-8}$ m^2/s, $K = 90 \pm 26$ KPa, $G = 66 \pm 19$ KPa, and $\kappa = 8.8 \times 10^{-13} \pm 5.6 \times 10^{-13}$ m^4/Ns, respectively. Some of these values were reported by Gu et al. [35], within order-of-magnitude differences, although our values included large errors.

In addition, the shear modulus G was in agreement with that measured in our oscillatory shear experiments.

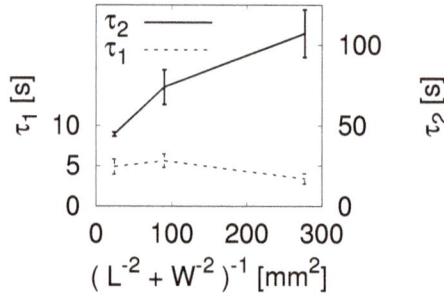

Figure 1. Relaxation times τ_1 (left axis) and τ_2 (right axis) as a function of $(L^{-2} + W^{-2})^{-1}$.

As we discuss in the following subsections, we conducted reciprocating friction experiments between a hemi-cylindrical indenter and an agarose hydrogel block at the sliding speed $V = 9.1$ mm/s and the stroke $S = 20$ mm under the normal load $F_N \approx 5$ N. By assuming the Hertzian contact [36], we could calculate the contact length as $L_c = \sqrt{4F_N R/\pi E^* W} \approx 0.003$ m ($R = 0.0145$ m, $E^* \approx 280$ KPa, and $W = 0.032$ m being the equivalent curvature radius, the equivalent elastic modulus, and the sample width, respectively) and the maximum pressure as $P_{max} = \sqrt{E^* F_N/(\pi RW)} \approx 30$ KPa. From these, we discuss the time scales for the contact: the contact time was $t_c = L_c/V \approx 0.3$ s, and the recurrence interval of the contact was $T = 2S/V \approx 4$ s. In addition, the characteristic time for water to be squeezed from the contact region was given by $t_{sq} \approx L_c^2/(\pi^2 D_c) \approx 7$ s. That is, the characteristic contact time was much shorter than the relaxation times due to water diffusion and viscoelasticity, and also much shorter than the recurrence time of the contact. This indicates that the agarose hydrogel could be regarded as an incompressible elastic material during our reciprocating friction experiments. In terms of stress, the maximum contact pressure ($P_{max} = 30$ KPa) was well below the failure stress ($\approx$80 KPa) of the gel, meaning that the friction experiments were performed under moderate mechanical conditions.

2.2. Effects of Adhesion Strength between Gel and Bottom Glass Slide

We then studied the frictional behavior of the agarose hydrogels under three different adhesion-strength conditions: As received (weak; for further details refer to the experimental section), Piranha treatment (intermediate), and Filter paper (strong). In this experiment, the normal load F_N was fixed as 4.41 N. Figure 2 shows time evolutions of the friction forces. The main figure and its inset correspond to time evolutions over the initial 100 s and those until entire rupture of the samples, respectively. As can be seen clearly, the Filter paper sample showed the largest friction forces for most of the experimental periods as well as the longest fatigue life until the entire rupture, that is, when the sample broke into two pieces. In contrast, the As received sample where the gel is adhered weakly to the glass slide broke less than 10% of the fatigue life of the gel fixed with the filter paper. For the samples with Piranha treatment, both the friction force and the fatigue life were in the middle of those values for the other two types of samples. This means that the stronger the adhesion between the gel and the glass slide, the longer the fatigue life becomes.

In order to understand the mechanisms, we analyzed images taken by a video camera during the friction experiments. Figure 3a,b shows the bottom views of the As received sample, and Figure 3d,e shows the side views of the Filter paper sample. It is clearly seen that, for the As received samples, a main crack was generated at the edge of the hydrogel surface and then propagated laterally in the direction perpendicular to the sliding direction. Figure 3c shows an image of the fracture surface after the friction experiment. Agarose gels behave as brittle materials, and surface cracks rather than internal cracks tend to be formed in unlubricated sliding contact. The striations, which were considered to be generated as a result of zig-zag propagation of fatigue cracks due to two different principal stress directions in reciprocating motions [37], indicated that a fatigue crack was formed at the frictional interface and propagated in the depth direction followed by the direction perpendicular to the sliding direction. On the other hand, for the Filter paper samples, a crack was formed at the edge of the sample in the same manner as for the As received sample but (as clearly seen in the fracture surface in Figure 3f) propagated slowly into the bottom, instead of following the progressive penetration and lateral propagation observed for the As received sample. It is important to note that some amount of wear debris is formed and accumulated on the frictional surface.

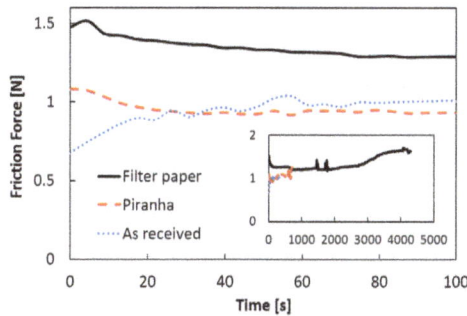

Figure 2. Time evolutions of friction forces for three samples of different adhesion strengths with bottom glass slide: Filter paper (solid line), Piranha (dashed line), and As received (dotted line). Main figure and inset correspond to the time evolutions of the initial 100 s and those until entire rupture of the samples, respectively. The fatigue lives were 4304 ± 56, 1457 ± 535, and 535 ± 288 s for Filter paper, Piranha, and As received samples, respectively. $F_N = 4.41$ N for all samples.

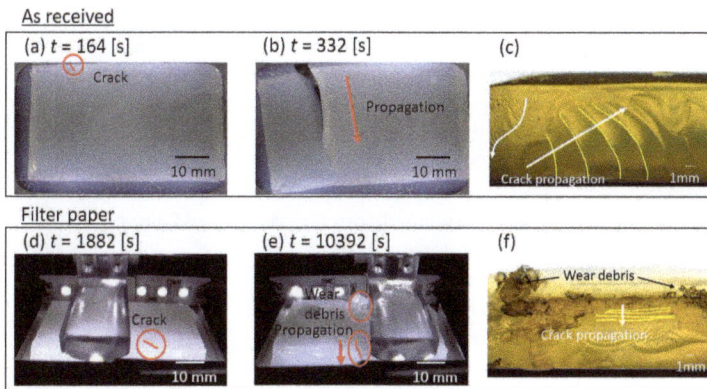

Figure 3. Fatigue crack behavior in As received (**a–c**) and Filter paper samples (**d–f**). (**a,b**) Bottom views, (**d,e**) side views, and (**c,f**) fracture surfaces after entire rupture of the samples. Small cracks are highlighted with red lines in (**a,d,e**). Arrows in (**c,f**) are crack paths, and yellow lines are supplementally drawn on striation patterns.

Figure 4 schematically depicts the experimentally observed behavior. For the As received samples, after an initial crack was formed at the edge of the frictional interface, it propagated in the depth direction, as shown in Figure 4a. Once it reached the interface between the gel and the glass slide (Figure 4b), it caused sliding at the bottom gel–glass interface and also enhanced tensile deformation of the gel in the sliding direction. This large tensile deformation accelerated fatigue crack propagation in the direction perpendicular to the sliding direction (Figure 4c), leading to fast rupture. On the other hand, for the Filter paper samples, the fatigue crack propagation slowed down because large tensile deformation to continue crack propagation was suppressed as a result of the strong adhesion between the gel and the bottom glass slide (Figure 4d,e). Along with the slow crack propagation, multiple fatigue cracks were successively formed, and the coalescence of two cracks occurred, leading to wear debris formation, as shown in Figure 4e,f.

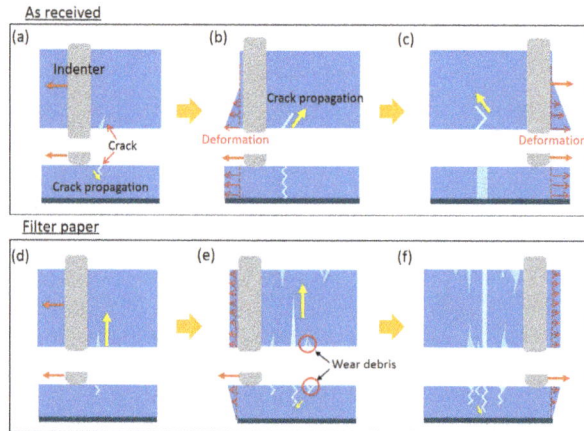

Figure 4. Schematic of propagation mechanisms of fatigue cracks and formation mechanisms of wear debris for (**a–c**) As received and (**d–f**) Filter paper samples.

We also discuss the mechanisms responsible for generating different friction forces for the three different samples at the initial stages of the friction experiments. Because the adhesion at the gel–glass interface was strong enough for the Filter paper samples (the measured maximum shear force was 13.8 N where the indenter–gel interface was ruptured; refer to the experimental section), sliding at the bottom interface did not occur during the friction experiments, and friction forces at the indenter–gel interface were measured. On the other hand, for the As received and Piranha samples with the maximum shear forces of 5.0 and 7.8 N, respectively, the adhesion at the bottom interface was not so strong and sliding started to occur at weaker forces. These led to different friction forces measured in the friction experiments and were consistent with our observations of frictional and fracture surfaces, showing some evidence of slip at the bottom interface.

2.3. Effects of Normal Load

We studied the frictional behavior under two different normal load conditions. In this experiment, the adhesion condition of the bottom interface was the Piranha treatment, giving intermediate adhesion strength. The main figure and the inset of Figure 5 correspond respectively to the time evolutions of the friction forces in the initial 100 s and those until the entire rupture of the sample. As could be expected, a larger friction force was measured (roughly proportional to the normal load, i.e., giving comparable friction coefficient) and shorter fatigue life was observed for a higher normal load. Figure 6 shows bottom views just before the entire rupture of the samples (Figure 6a at $t = 6899$ s for $F_N = 2.89$ N, and Figure 6c at 801 s for 5.89 N), and their fracture surfaces (Figure 6b,d, respectively). It is seen that a greater number of cracks were generated and slower crack propagation in the depth directions was observed for a smaller normal load. Moreover, a large amount of wear debris was formed only for the smaller normal load.

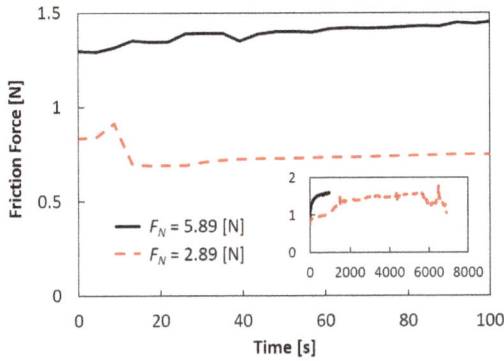

Figure 5. Time evolutions of friction forces under two different normal load conditions: F_N of 5.89 N (solid line) and 2.89 N (dashed line). Main figure and inset correspond to the time evolutions of the initial 100 s and those until entire rupture of the samples. The fatigue lives were 1542 ± 938 and 4559 ± 1688 s for F_N of 5.89 and 2.89 N, respectively. Piranha treatment was used for all samples.

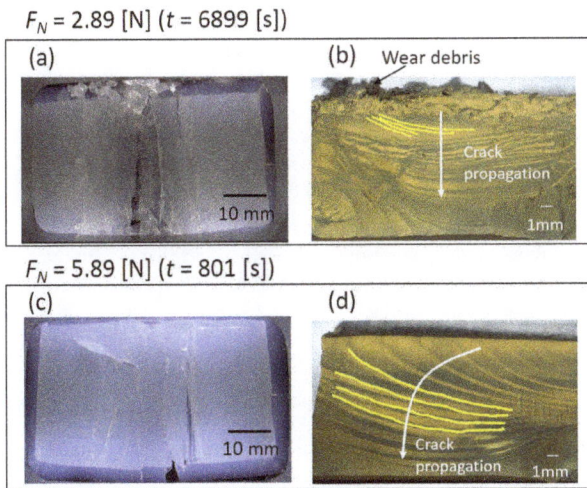

Figure 6. Fatigue crack behavior under two different normal load conditions: (**a,b**) $F_N = 2.89$ N, and (**c,d**) $F_N = 5.89$ N. Images from bottoms views (**a,c**) are taken just before entire rupture of each sample.

The mechanisms for generating such differences are explained in Figure 7. For smaller normal load, a smaller friction force is generated. As a result, a weaker driving force for fatigue crack propagation is applied, leading to slower crack propagation and a longer fatigue life. This gives enough time for creating new cracks (Figure 7b), and if two cracks happen to merge, enclosed regions are detached from the bulk of the gel and are ejected out as wear debris. On the other hand, if the normal load is large, the fatigue cracks propagate fast enough to reach the bottom interface without creating wear debris.

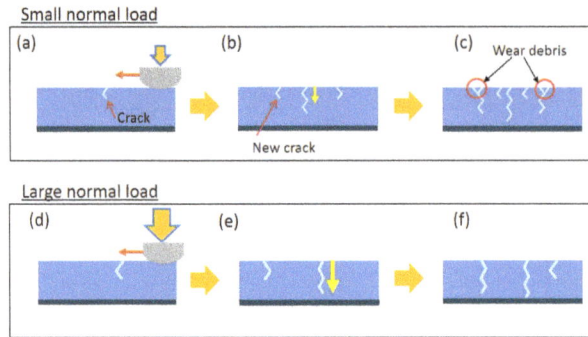

Figure 7. Schematic of propagation mechanisms of fatigue cracks and formation mechanisms of wear debris (**a–c**) for small normal loads and (**d–f**) for large normal loads. We note that the crack speed, i.e., time to rupture, was different between these two conditions.

2.4. Toward the Toughening of Hydrogels

On the basis of our results, two important points can be inferred to improve the toughness of hydrogels as frictional materials. One point is that the fixation of the gel has to be made in an appropriate manner. This leads to the slowing down of the fatigue cracks in the depth direction and a longer fatigue life, as discussed in Section 2.2. Another important point is the reinforcement of the gel along the sliding direction to support generated tensile stress and to avoid large deformation of the gel. For this purpose, the insertion of fibers into the gel matrix would be one option, as examined by Sakai and co-workers to reduce the friction of poly(vinyl alcohol) (PVA) hydrogels [38]. This is also expected to contribute to the toughening of hydrogels against friction. Both will be interesting and important topics for future studies.

3. Conclusions

We studied the propagation behavior of fatigue cracks during reciprocating friction experiments between a PMMA indenter and an agarose hydrogel. We found that the friction force and fatigue life were strongly influenced by the applied normal load as well as the adhesion strength at the bottom glass–gel interface. We observed the propagation speed and path of the fatigue cracks, both of which were also affected by these conditions.

4. Experiment

4.1. Sample

We prepared 3 wt % agarose hydrogels on glass slides. After dissolving agarose powder (Agarose III, Wako Pure Chemical Industries, Osaka, Japan) into hot water and stirring with an agitator, we poured the solution onto a glass slide with a rubber mold and solidified the sample in a refrigerator at 4 °C for 12 h. The sample size was L (length along the sliding direction) = 56 mm, W (width) = 32 mm, and H (thickness) = 10 mm. In this study, in order to examine the effects of adhesion between the gel and the bottom glass slide, we treated the glass surfaces in three different manners: one as received (hereafter called "As received", leading to weak adhesion between the gel and glass), one with piranha treatment ("Piranha"; intermediate adhesion), and one by gluing filter paper on the glass surface ("Filter paper"; strong adhesion due to penetration of agarose solution into the filter paper before gelation). When the piranha treatment was made, a glass slide was soaked into a mixture of 30 wt % hydrogen peroxide solution (Wako Pure Chemical Industries, Japan) to sulfuric acid (Wako Pure Chemical Industries, Japan) in a ratio of 1:3 (by weight) at 80 °C for 1 h before pouring the agarose solution.

4.2. Friction Experiment

A schematic of our experimental setup is shown in Figure 8. We conducted reciprocating friction experiments between an agarose hydrogel and a hemi-cylindrical indenter made of polymethyl methacrylate (PMMA). The curvature radius R and length L of the indenter were 14.5 and 57 mm, respectively. The indenter was placed in the direction perpendicular to the sliding direction so that it crossed both side edges of the gel sample. The sliding speed V was 9.1 mm/s, and the stroke S was 20 mm. The applied normal loads F_N were 2.94, 4.41, and 5.89 N. We measured lateral forces acting on the bottom plate with a load cell at 100 Hz. We took the largest and smallest (with negative sign) 10 points of the friction forces per cycle, calculated average over absolute values of these 20 points, and represented it as a friction force per cycle. Under each condition, we repeated the friction experiments three times. All hydrogel samples were used as prepared, and all experiments were performed at room temperature under unlubricated conditions (without supplying additional water) with the aim to continue the experiments in a controlled manner.

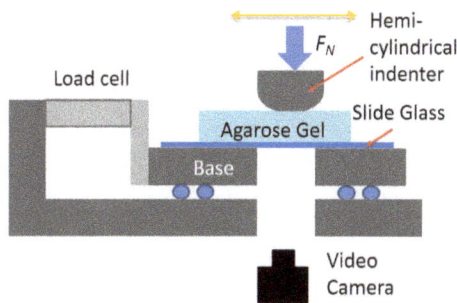

Figure 8. Schematic of reciprocating friction experiments.

4.3. Visualization

We observed crack propagation behavior during the friction experiments with a video camera (GZ-G5, JVC KENWOOD, Yokohama, Japan) from the bottom of the glass slide, as shown in Figure 8. Only when Filter paper samples were tested (bottom view was not available because of an opaque filter paper adhered on the glass) was observation made from the side.

4.4. Characterization of Mechanical Properties

Because hydrogels are soft solids composed of sparsely cross-linked polymers and a large amount of water, they exhibit mechanical relaxation due to viscoelasticity of polymer chains and diffusive transport of water molecules inside the polymer network (collective diffusion). In order to characterize such relaxation behavior, we conducted oscillatory shear experiments and unconfined stress relaxation experiments. In addition, we also conducted uni-axial compression experiments to determine the failure stress.

In the oscillatory shear experiments, we prepared cylindrical hydrogel samples ($\phi = 25$ mm, $t = 5$ mm) and applied the shear of the strain amplitude $\gamma = 0.01$ and of the frequency $f = 0.001$–10 Hz with a parallel plate ($\phi = 25$ mm) using a rheometer (MCR-301, Anton Paal, Graz, Austria). Figure 9a shows the frequency dependencies of the storage modulus $G'(f)$ and the loss modulus $G''(f)$. The shear modulus estimated from G' was about 70 KPa. On the other hand, a clear peak in $G''(f)$ was seen around 0.03 Hz. This indicates that there existed a viscoelastic relaxation mode whose characteristic time was about 30 s.

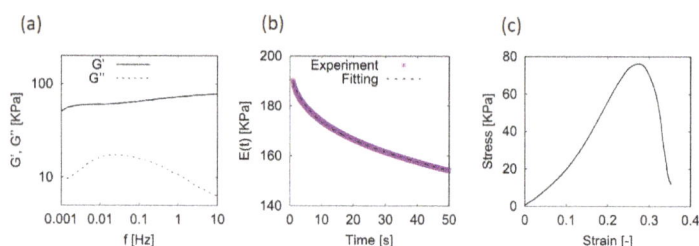

Figure 9. (**a**) Linear viscoelastic moduli $G'(f)$ (storage modulus) and $G''(f)$ (loss modulus) at room temperature, (**b**) typical example of stress-relaxation behavior ($l = 25$ mm, $w = 10$ mm, $h = 5$ mm), and (**c**) stress–strain curve for the uni-axial compression test.

In the stress-relaxation experiments, we prepared rectangular blocks with three different widths ($L = 25$ mm; $W = 5, 10, 20$ mm; $H = 5$ mm). We sandwiched each sample between two glass plates, applied unconfined uni-axial compression at $V = 1$ mm/s, and then fixed the displacement δ around 0.3 mm while measuring the normal force $F(t)$ as a function of time. A typical relaxation behavior from $t = 1$ to 50 s is shown in Figure 9b. The relaxation function $E(t) = \sigma_z(t)/\epsilon_z = F(t)/(LW)H/\delta$ (σ_z and ϵ_z being the normal stress and the compressive strain, respectively) seemed to have two (short and long) relaxation modes; thus we fitted the experimental data with the following equation:

$$E(t) = E_\infty + E_1 exp(-\frac{t}{\tau_1}) + E_2 exp(-\frac{t}{\tau_2}), \tag{6}$$

where E_∞ is the relaxed modulus, and E_1 (E_2) and τ_1 (τ_2) are, respectively, the intensity and the characteristic time of the shorter (longer) relaxation mode. We applied the least-squares fitting with the grid-search technique to find optimum values for E_∞, E_1, E_2, τ_1, and τ_2. The obtained fitting curve is also plotted in Figure 9b and is in excellent agreement with the original curve.

In the uni-axial compression experiments, we prepared cylindrical hydrogel samples ($\phi = 10$ mm, $t = 5$ mm). We sandwiched a sample with two glass plates and measured the normal force during compression at $V = 0.1$ mm/s until it broke. Figure 9c is a typical result for the stress–strain curve, showing that the failure stress and strain were about 80 KPa and 0.28, respectively.

4.5. Evaluation of Adhesion Strength between Gel and Bottom Slide Glass

In order to evaluate the adhesion strength between the agarose hydrogel and glass slide, we prepared gel samples in the three different manners noted above. During the gelation process, we also contacted an indenter gluing filter paper on the top surface of the agarose solution in order to fix the gel firmly against the indenter. As a consequence, when a lateral displacement of $S = 20$ mm was applied at $V = 9.1$ mm/s and $F_N = 4.41$ N (in the same manner as the first cycle in the friction test), delamination occurred at the bottom glass–hydrogel interface for the As received and Piranha samples without any slip at the top indenter–hydrogel interface. Only when Filter paper samples were tested was fracture along the top interface observed. The measured maximum shear forces were 13.8, 7.8, and 5.0 N for the Filter paper, Piranha, and As received samples, respectively.

Author Contributions: Conceptualization, T.Y. and Y.S., Data Curation, R.S. and T.Y., Writing—Original Draft Preparation, R.S and T.Y., Writing—Review & Editing, Y.S., Visualization, T.Y., Supervision, T.Y., Funding Acquisition, T.Y.

Acknowledgments: The authors thank T. Tanaka and T. Morita for experimental support and S. K. Sinha for carefully reading the manuscript. This work was supported by JSPS KAKENHI Grant No. JP18K03565.

Conflicts of Interest: The authors declare no conflict of interest.

Appendix A

Here we consider a rectangular hydrogel block with length L, width W, and height H. As we discussed in the previous section, viscoelastic relaxation occurs much faster than that due to the diffusive transport of water (solvent) in agarose hydrogels. Thus we assume that the gel is made of an elastic network and a large amount of water.

The constitutive relation for the polymer network is described by the following equation:

$$\sigma_{\alpha\beta} = K\frac{\partial u_\gamma}{\partial r_\gamma}\delta_{\alpha\beta} + G\Big(\frac{\partial u_\alpha}{\partial r_\beta} + \frac{\partial u_\beta}{\partial r_\alpha} - \frac{2}{3}\frac{\partial u_\gamma}{\partial r_\gamma}\delta_{\alpha\beta}\Big), \tag{A1}$$

where K and G are the osmotic and shear moduli, respectively, and $\mathbf{u} = (u_x, u_y, u_z)$ is the elastic displacement.

By taking into account the diffusive transport of water, the force balances in terms of the polymer and water are described by the following equations:

$$\bar{\xi}(\dot{\mathbf{u}} - \mathbf{v}_s) = \nabla \cdot \sigma - \phi \nabla p, \tag{A2}$$

$$\bar{\xi}(\mathbf{v}_s - \dot{\mathbf{u}}) = -(1 - \phi)\nabla p, \tag{A3}$$

where $\bar{\xi}$ is the friction constant for the relative motion between the polymer and water; $\dot{\mathbf{u}}$ and $\mathbf{v}_s$ are the local velocity of the polymer and water, respectively; ϕ is the volume fraction of the polymer; and p is the local isotropic pressure.

On the other hand, $\dot{\mathbf{u}}$ and $\mathbf{v}_s$ must satisfy the incompressible condition:

$$\nabla \cdot \mathbf{v} = \nabla \cdot \{\phi\dot{\mathbf{u}} + (1 - \phi)\mathbf{v}_s\} = 0, \tag{A4}$$

where $\mathbf{v}$ is the mean velocity. Equation (A3) can be written with the mean velocity:

$$\dot{\mathbf{u}} - \mathbf{v} = \kappa_0(1 - \phi)\nabla p, \tag{A5}$$

where $\kappa_0 = 1/\bar{\xi}$. Substituting Equation (A5) into Equation (A4), we obtain

$$\frac{\partial \omega}{\partial t} = \kappa_0(1 - \phi)\nabla^2 p, \tag{A6}$$

where $\omega = \partial u_x/\partial x + \partial u_y/\partial y + \partial u_z/\partial z$ is the volumic strain of the polymer.

From the force balance equations (Equations (A2) and (A3)), we obtain the Poisson equation in terms of pressure:

$$\nabla^2 p = \Big(K + \frac{4}{3}G\Big)\nabla^2 \omega, \tag{A7}$$

and combining Equation (A7) with Equation (A6), we also obtain the diffusion equation:

$$\frac{\partial \omega}{\partial t} = D_c\nabla^2 \omega, \tag{A8}$$

where $D_c = \kappa_0(1 - \phi)(K + 4G/3)$ is the collective diffusion coefficient. Thus, the problem is to solve the diffusion equation (Equation (A8)) with appropriate boundary conditions.

We consider the situation in which we apply uni-axial compression along the z-direction. Here we assume elastic displacements as follows:

$$\begin{aligned} u_x &= u_x(x, y, t), \\ u_y &= u_y(x, y, t), \\ u_z &= \epsilon_z(t)z. \end{aligned} \tag{A9}$$

The stresses are described by

$$
\begin{aligned}
\sigma_{xx} &= (K - \frac{2}{3}G)\omega + 2G\frac{\partial u_x}{\partial x}, \\
\sigma_{yy} &= (K - \frac{2}{3}G)\omega + 2G\frac{\partial u_y}{\partial y}, \\
\sigma_{zz} &= (K - \frac{2}{3}G)\omega + 2G\epsilon_z.
\end{aligned}
\tag{A10}
$$

In an unconfined compression test, the total stresses along the x- and y-directions must equal zero at any point inside the gel. From these conditions,

$$
p(x,y,t) = (K - \frac{2}{3}G)\omega + 2G\frac{\partial u_x}{\partial x} = (K - \frac{2}{3}G)\omega + 2G\frac{\partial u_y}{\partial y},
\tag{A11}
$$

and we obtain $\partial u_x/\partial x = \partial u_y/\partial y$. After some simple calculations, Equation (A10) can be written in the following manner:

$$
\begin{aligned}
\sigma_{xx} &= \sigma_{yy} = (K + \frac{G}{3})\omega - G\epsilon_z, \\
\sigma_{zz} &= (K - \frac{2}{3}G)\omega + 2G\epsilon_z.
\end{aligned}
\tag{A12}
$$

Because the elastic stresses must vanish on the side faces of the gel and because the compressive strain is held fixed at $t = 0$ as $\epsilon_z(t) = \epsilon_{z0}$, we obtain the following equations:

$$
\begin{aligned}
\omega(0,y,t) = \omega(L,y,t) &= \frac{G}{K+G/3}\epsilon_{z0}, \\
\omega(x,0,t) = \omega(x,W,t) &= \frac{G}{K+G/3}\epsilon_{z0}.
\end{aligned}
\tag{A13}
$$

Furthermore, the pressure must vanish at $t = \infty$, and thus

$$
\omega(x,y,\infty) = \omega_{eq} = \frac{G}{K+G/3}\epsilon_{z0}.
\tag{A14}
$$

Now we are ready to solve Equation (A8). We assume the solution as $\omega(x,y,t) = \omega_{eq} + g(x,y)exp(-t/\tau)$ and then substitute it into Equation (A8). We obtain

$$
\nabla^2 g(x,y) = -\frac{1}{D_c\tau}g(x,y).
\tag{A15}
$$

By considering the boundary conditions (Equations (A13) and (A14)), we obtain the following solution:

$$
\omega(x,y,t) = \frac{G}{K+G/3}\epsilon_{z0}\{1 - \frac{\pi^2}{4}sin(\frac{\pi x}{L})sin(\frac{\pi y}{W})exp(-\frac{t}{\tau})\},
\tag{A16}
$$

where we take the lowest-order term only. The relaxation time τ is given by

$$
\tau = \frac{1}{D_c}\frac{1}{\pi^2(\frac{1}{L^2} + \frac{1}{W^2})}.
\tag{A17}
$$

This indicates that the relaxation time is proportional to the square of the sample size (e.g., if $L \gg W$, then $\tau \propto W^2$). Finally, the expression for the average compressive stress $\overline{\sigma_{zz}}$ is obtained:

$$\overline{\sigma_{zz}}(t) = \frac{G}{K + G/3} \epsilon_{z0} \{3K + G exp(-\frac{t}{\tau})\}. \tag{A18}$$

References

1. Gong, J.P. Friction and lubrication of hydrogels—Its richness and complexity. *Soft Matter* **2006**, *2*, 544–552. [CrossRef]
2. Kaneko, D.; Tada, T.; Kurokawa, T.; Gong, J.P.; Osada, Y. Mechanically Strong Hydrogels with Ultra-Low Frictional Coefficients. *Adv. Mater.* **2005**, *17*, 535–538. [CrossRef]
3. Gong, J.P.; Kurokawa, T.; Narita, T.; Kagata, G.; Osada, Y.; Nishimura, G.; Kinjo, M. Synthesis of Hydrogels with Extremely Low Surface Friction. *J. Am. Chem. Soc.* **2001**, *123*, 5582–5583. [CrossRef] [PubMed]
4. Takata, M.; Yamaguchi, T.; Doi, M. Electric Field Effect on the Sliding Friction of a Charged Gel. *J. Phys. Soc. Jpn.* **2009**, *78*, 084602. [CrossRef]
5. Takata, M.; Yamaguchi, T.; Doi, M. Friction Control of a Gel by Electric Field in Ionic Surfactant Solution. *J. Phys. Soc. Jpn.* **2010**, *79*, 063602. [CrossRef]
6. Suzuki, R.; Yamaguchi, T.; Doi, M. Frictional Property of Hydrogels Prepared under Electric Fields. *J. Phys. Soc. Jpn.* **2013**, *82*, 124803. [CrossRef]
7. Jin, Z.; Dowson, D. Bio-friction. *Friction* **2013**, *1*, 100–113. [CrossRef]
8. Murakami, T.; Higaki, H.; Sawae, Y.; Ohtsuki, N.; Moriyama, S.; Nakanishi, Y. Adaptive multimode lubrication in natural synovial joints and artificial joints. *Proc. IMechE H J. Eng. Med.* **1998**, *212*, 23–35. [CrossRef] [PubMed]
9. Freeman, M.E.; Furey, M.J.; Love, B.J.; Hampton, J.M. Friction, wear, and lubrication of hydrogels as synthetic articular cartilage. *Wear* **2000**, *241*, 129–135. [CrossRef]
10. Murakami, T.; Yarimitsu, S.; Nakashima, K.; Yamaguchi, T.; Sawae, Y.; Sakai, N.; Suzuki, A. Superior lubricity in articular cartilage and artificial hydrogel cartilage. *Proc. IMechE J J. Eng. Tribol.* **2014**, *228*, 1099–1111. [CrossRef]
11. Murakami, T.; Sakai, N.; Yamaguchi, T.; Yarimitsu, S.; Nakashima, K.; Sawae, Y.; Suzuki, A. Evaluation of a superior lubrication mechanism with biphasic hydrogels for artificial cartilage. *Tribol. Int.* **2015**, *89*, 19–26. [CrossRef]
12. Bray, J.C.; Merrill, E.W. Poly(vinyl alcohol) hydrogels for synthetic articular cartilage material. *J. Biomed. Mat. Res.* **1973**, *7*, 431–443. [CrossRef] [PubMed]
13. Gu, Z.Q.; Xiao, J.M.; Zhang, X.H. The development of artificial articular cartilage—PVA-hydrogel. *Bio-Med Mater. Eng.* **1998**, *8*, 75–81.
14. Stammen, J.A.; Williams, S.; Ku, D.N.; Guldberg, R.E. Mechanical properties of a novel PVA hydrogel in shear and unconfined compression. *Biomaterials* **2001**, *22*, 799–806. [CrossRef]
15. Pan, Y.; Xiong, D. Friction properties of nano-hydroxyapatite reinforced poly(vinyl alcohol) gel composites as an articular cartilage. *Wear* **2009**, *266*, 699–703. [CrossRef]
16. Murakami, T.; Yarimitsu, S.; Nakashima, K.; Sakai, N.; Yamaguchi, T.; Sawae, Y.; Suzuki, A. Biphasic and boundary lubrication mechanisms in artificial hydrogel cartilage: A review. *Proc. IMechE H J. Eng. Med.* **2015**, *229*, 864–878. [CrossRef] [PubMed]
17. Murakami, T.; Yarimitsu, S.; Sakai, N.; Nakashima, N.; Yamaguchi, T.; Sawae, Y. Importance of adaptive multimode lubrication mechanism in natural synovial joints. *Tribol. Int.* **2017**, *113*, 306–315. [CrossRef]
18. Murakami, T.; Yarimitsu, S.; Sakai, N.; Nakashima, K.; Yamaguchi, T.; Sawae, Y.; Suzuki, A. Superior lubrication mechanism in poly (vinyl alcohol) hybrid gel as artificial cartilage. *Proc. IMechE J J. Eng. Tribol.* **2017**, *231*, 1160–1170. [CrossRef]
19. Gong, J.P.; Katsuyama, Y.; Kurokawa, T.; Osada, Y. Double-Network Hydrogels with Extremely High Mechanical Strength. *Adv. Mater.* **2003**, *15*, 1155–1158. [CrossRef]
20. Yasuda, K.; Gong, J.P.; Katsuyama, Y.; Nakayama, A.; Tanabe, Y.; Kondo, E.; Ueno, M.; Osada, Y. Biomechanical properties of high-toughness double network hydrogels. *Biomaterials* **2005**, *26*, 4468–4475. [CrossRef] [PubMed]

21. Haque, M.A.; Kurokawa, T.; Kamita, G.; Gong, J.P. Lamellar Bilayers as Reversible Sacrificial Bonds To Toughen Hydrogel: Hysteresis, Self-Recovery, Fatigue Resistance, and Crack Blunting. *Macromolecules* **2011**, *44*, 8916–8924. [CrossRef]
22. Kundu, S.; Crosby, A.J. Cavitation and fracture behavior of polyacrylamide hydrogels. *Soft Matter* **2009**, *5*, 3963–3968. [CrossRef]
23. Boue, T.G.; Harpaz, R.; Fineberg, J.; Bouchbinder, E. Failing softly: A fracture theory of highly-deformable materials. *Soft Matter* **2015**, *11*, 3812–3821. [CrossRef] [PubMed]
24. Zhao, X. Multi-scale multi-mechanism design of tough hydrogels: Building dissipation into stretchy networks. *Soft Matter* **2014**, *10*, 672–687. [CrossRef] [PubMed]
25. Liu, T.; Long, R.; Hui, C.-Y. The energy release rate of a pressurized crack in soft elastic materials: Effects of surface tension and large deformation. *Soft Matter* **2014**, *10*, 7723–7729. [CrossRef] [PubMed]
26. Liu, T.; Jagota, A.; Hui, C.-Y. Adhesive contact of a rigid circular cylinder to a soft elastic substrate—The role of surface tension. *Soft Matter* **2015**, *11*, 3844–3851. [CrossRef] [PubMed]
27. Baumberger, T.; Caroli, C.; Martina, D.; Ronsin, O. Magic angles and cross-hatching instability in hydrogel fracture. *Phys. Rev. Lett.* **2008**, *100*, 1–4, doi:10.1103/PhysRevLett.100.178303. [CrossRef] [PubMed]
28. Tanaka, Y.; Shimazaki, R.; Yano, S.; Yoshida, G.; Yamaguchi, T. Solvent effects on the fracture of chemically crosslinked gels. *Soft Matter* **2016**, *12*, 8135–8142. [CrossRef] [PubMed]
29. Suciu, A.N.; Iwatsubo, T.; Matsuda, M.; Nishino, T. A Study upon Durability of the Artificial Knee Joint with PVA Hydrogel Cartilage. *JSME Int. J. C* **2004**, *47*, 199–208. [CrossRef]
30. Doi, M. *Soft Matter Physics*; Oxford University Press: Oxford, UK, 2013.
31. Mow, V.C.; Kuei, S.C.; Lai, W.M.; Armstrong, C.G. Biphasic Creep and Stress Relaxation of Articular Cartilage in Compression: Theory and Experiments. *J. Biomech. Eng.* **1980**, *102*, 73–84. [CrossRef] [PubMed]
32. Ateshian, G.A.; Lai, W.M.; Zhu, W.B.; Mow, V.C. An asymptotic solution for the contact of two biphasic cartilage layers. *J. Biomech.* **1994**, *27*, 1347–1360. [CrossRef]
33. Ateshian, G.A. The Role of Interstitial Fluid Pressurization in Articular Cartilage Lubrication. *J. Biomech.* **2009**, *42*, 1163–1176. [CrossRef] [PubMed]
34. Tanaka, T.; Fillmore, D.J. Kinetics of swelling of gels. *J. Chem. Phys.* **1979**, *70*, 1214. [CrossRef]
35. Gu, W.Y.; Yao, H.; Huang, C.Y.; Cheung, H.S. New insight into deformation-dependent hydraulic permeability of gels and cartilage, and dynamic behavior of agarose gels in confined compression. *J. Biomech.* **2003**, *36*, 593–598. [CrossRef]
36. Johnson, K.L. *Contact Mechanics*; Cambridge University Press: Cambridge, UK, 2003.
37. Cotterell, B.; Rice, J.R. Slightly curved or kinked cracks. *Int. J. Fract.* **1980**, *16*, 155–169. [CrossRef]
38. Sakai, N.; Hashimoto, C.; Yarimitsu, S.; Sawae, Y.; Komori, M.; Murakami, T. A functional effect of the superficial mechanical properties of articular cartilage as a load bearing system in a sliding condition. *Biosurf. Biotribol.* **2016**, *2*, 26–39. [CrossRef]

gels

MDPI

Article

Effect of Gamma Ray Irradiation on Friction Property of Poly(vinyl alcohol) Cast-Drying on Freeze-Thawed Hybrid Gel

Saori Sasaki [1,2,*], Seiji Omata [1,3], Teruo Murakami [1,4], Naotsugu Nagasawa [5], Mitsumasa Taguchi [5] and Atsushi Suzuki [6]

[1]   Research Center for Advanced Biomechanics, Kyushu University, 744 Motooka, Nishi-ku, Fukuoka 819-0395, Japan; s-omata@mech.nagoya-u.ac.jp (S.O.); tmura@mech.kyushu-u.ac.jp (T.M.)
[2]   Institute for Material Chemistry and Engineering, Kyushu University, 744 Motooka, Nishi-ku, Fukuoka 819-0395, Japan
[3]   Department of Micro-Nano Mechanical Science and Engineering, Nagoya University, Furo-cho, Chikusa-ku, Nagoya 464-8603, Japan
[4]   Faculty of Fukuoka Medical Technology, Teikyo University, 6-22 Misaki-machi, Omuta 836-8505, Japan
[5]   Takasaki Advanced Radiation Research Institute, National Institutes for Quantum and Radiological Science and Technology, Watanuki 1233, Takasaki, Gunma 370-1292, Japan; nagasawa.naotsugu@qst.go.jp (N.N.); taguchi.mitsumasa@qst.go.jp (M.T.)
[6]   Research Institute of Environment and Information Sciences, Yokohama National University, 79-7 Tokiwadai, Hodogaya-ku, Yokohama 240-8501, Japan; asuzuki@ynu.ac.jp
*   Correspondence: sasaki.saori.004@m.kyushu-u.ac.jp; Tel.: +81-92-802-2508

Received: 27 February 2018; Accepted: 27 March 2018; Published: 29 March 2018

Abstract: Poly(vinyl alcohol) (PVA) is a biocompatible polymer with low toxicity. It is possible to prepare physically cross-linked PVA gels having hydrogen bonds without using a cross-linking agent. The newly reported physically cross-linked PVA cast-drying (CD) on freeze-thawed (FT) hybrid gel has an excellent friction property, which is expected to be applied as a candidate material for artificial cartilage. Gamma ray sterilization for clinical applications usually causes additional chemical cross-linking and changes physical properties of gels. In this study, CD on FT hybrid gels were irradiated using gamma rays at a different dose rate and irradiance. The results showed the optimized irradiation conditions for gamma irradiated gels to retain excellent friction characteristics.

Keywords: PVA gel; gamma ray sterilization; artificial hydrogel cartilage; frictional property; wear

1. Introduction

Poly(vinyl alcohol) (PVA) is a high biocompatible synthetic polymer with low toxicity and excellent mechanical properties. Physically cross-linked PVA gels having hydrogen bonds are prepared without using a cross-linking agent via two methods. One is the repeated freeze-thawing (FT) of aqueous PVA solutions [1], while the other one is the cast-drying (CD) of them [2,3]. These two physically cross-linked PVA gels have three-dimensional amorphous network structures which are physically cross-linked by microcrystallites. The distributions of microcrystallites, however, are quite different. While FT gel has a heterogeneous network structure with high permeability, CD gel has a uniform network structure with low permeability [4]. The possibility for application of FT gels is already reported as candidate materials for artificial cartilage, such as hip and knee prostheses [5–7]. It is found that FT gels show a good frictional property similar to the natural articular cartilage. On the other hand, CD gels reduce the friction level compared with FT gels and the natural articular cartilage in reciprocating friction tests in saline solution [8,9]. The coefficient of friction of CD gel in saline solution was about 0.05, which was superior to FT gel (0.20) and natural articular cartilage (0.15) [9].

Recently, a method for preparing the physically cross-linked PVA gel as PVA CD on FT hybrid gel was newly reported by Suzuki et al. [10]. The CD gel was laminated on the FT gel. This PVA CD on FT hybrid gel presents excellent friction properties in reciprocating friction tests in water and physiological saline [4,10,11]. The coefficient of friction of PVA CD on FT hybrid gel was in a lower range than 0.005 and the wear was minimal in the simulated synovial fluid, whereas that of FT and CD gel reached 0.012 and 0.01, respectively [12]. Therefore, such PVA CD on FT hybrid gel is expected to be applied as a candidate material for artificial cartilage.

In clinical applications, sterilization treatment is a mandatory procedure. There are several methods for sterilization. Autoclave sterilization is a widely used one in medical fields. The microorganisms are killed by heating with saturated steam at an appropriate temperature over 121 °C in a sealed apparatus. The advantage of this method is that heat can penetrate quickly into the deep part of the object and surely kill microorganisms in a short time. Therefore, this method is widely used in medical institutions and pharmaceutical manufacturing sites. However, heating using saturated steam destroys the microcrystallines as cross-linked regions. Therefore, autoclave sterilization cannot be used for physically cross-linked PVA gels. Besides, in a previous study, we soaked a hybrid gel into 70% ethanol solution for sterilization for animal experiments. The friction property of the CD on FT hybrid gel, however, was deteriorated due to shrinkage of gel surface [13]. Therefore, gamma ray sterilization is considered as a suitable method. There are many reports on gamma ray sterilization of gels [14,15]. Moderate gamma ray sterilization does not destroy microcrystals, and the gel shape is possible to maintain. Nevertheless, additional chemical cross-linking could occur, so the physical properties of PVA CD on FT hybrid gels are expected to change after gamma ray irradiation [16–18]. Therefore, it is necessary to investigate the effect of gamma rays on frictional property of PVA CD on FT hybrid gels.

In this study, we irradiated the PVA CD on FT hybrid gels under different conditions, including varied irradiation rate and total irradiance, to investigate the relationship between gamma ray irradiation and frictional properties. Then, friction and wear tests were performed in a reciprocating friction test at two lubricant temperatures, i.e., room temperature and body temperature (37 °C).

2. Results and Discussion

We irradiated the gels using gamma rays under various conditions of irradiation rate and total irradiance as described in Section 4.1.2 (Gamma Ray-Irradiation). In order to evaluate the tribological properties for artificial cartilage, frictional properties were characterized by the reciprocating friction test in this study. In order to evaluate the wear and durability, the sample surface was observed using a phase contrast microscope after completion of the reciprocating friction test. From the obtained results, we investigated the changes of friction and wear characteristics of the gels irradiated using gamma rays and achieved the optimum irradiation conditions.

2.1. Swelling Property

One of the characteristics of PVA gels is a water content as high as about 70–80%, which is close to that of articular cartilage in human body (about 70–80%) [19]. The change of water content before and after gamma ray irradiation was investigated. The result is shown in Figure 1. Before gamma ray irradiation, the water content of PVA CD on FT hybrid gels reached 75%. All samples after gamma ray irradiation showed lower water contents than the ones before irradiation. The reduction in the water contents was caused by the process in which additional chemically cross-linking was introduced during gamma ray irradiation. The water content tended to decrease slightly as the irradiance of gamma rays increased.

Although the water content decrement occurred after irradiation using gamma rays, it was maintained around 70%, which is close to that of articular cartilage. Therefore, the PVA CD on FT hybrid gel after gamma ray irradiation can be potentially used for artificial cartilage with biphasic property. It is considered that the layered structure composing of CD and FT gel layers with appropriate microcrystallites maintained the necessary water content. Further investigations to elucidate the swelling behaviors after gamma ray irradiation are required.

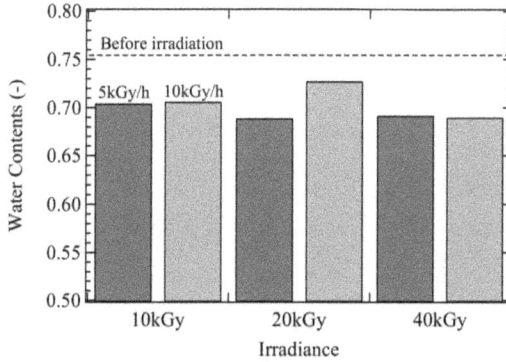

Figure 1. Water contents of poly(vinyl alcohol) PVA cast-drying (CD) on freeze-thawed (FT) hybrid gels before and after gamma ray irradiation. The broken line shows the water content of the gel before gamma irradiation.

2.2. Frictional Properties

2.2.1. Reciprocating Test at Room Temperature

To investigate the effect of gamma ray irradiation on frictional properties, the reciprocating test at room temperature was conducted first. The changes in coefficient of friction, μ_k, are shown in Figure 2. The samples as shown in Figure 2a,b were prepared at the dose rate of 5 and 10 kGy/h, respectively. At the dose rate of 5 kGy/h, μ_k for 10 kGy irradiance was noticed to be reduced from non-irradiation. To compare friction levels among irradiated gels, μ_k increased as the irradiance increased at each dose rate. In addition, it is also noticed that the increase in μ_k depended on the dose rate. At the same irradiance, the gel at a high dose rate showed a high μ_k except at 40 kGy.

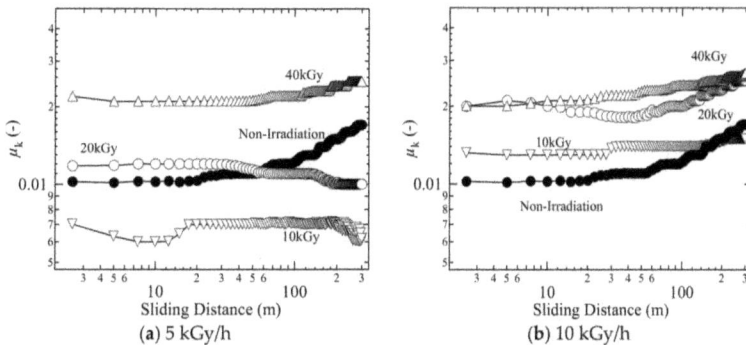

Figure 2. Coefficient of friction of gamma ray irradiated gels with (**a**) 5 kGy/h and (**b**) 10 kGy/h in reciprocating friction tests at room temperature; close circle, non-irradiation gel; inverted triangle, irradiance = 10 kGy; open circle, irradiance = 20 kGy; triangle, irradiance = 40 kGy.

The changes in μ_k with irradiation are considered to be related to the variation in shearing resistance of the irradiated gels. As the surface solidifies due to the increase in chemical cross-linking points, the shear strength/resistance increases, which increases the μ_k. However, the reduction in friction for 10 kGy at 5 kGy/h from non-irradiation level appeared to be related to the improvement of shearing strength to frictional force. The high friction at high dose rate suggests that the surface physical properties intensely changed by receiving a lot of gamma rays in a short time. The size and distribution of cross-linking points on the surface changed, resulting from the variation in the amount of oxygen around the gel at the time of irradiation with gamma rays. Therefore, in order to maintain the friction characteristics of PVA CD on FT hybrid gels, it is important to reduce the dose rate and irradiance.

2.2.2. Reciprocating Test at 37 °C

To investigate the frictional property of gamma ray irradiation gels in an environment close to human body, the temperature of the lubricating liquid was raised at 37 °C. The results are shown in Figure 3.

The samples as shown in Figure 3a,b were prepared at dose rate 5 and 10 kGy/h, respectively. Compared with the result at room temperature (Figure 2), the non-irradiation gel showed a higher μ_k at 37 °C. On the other hand, the gamma ray irradiated gel at 37 °C did not show evident changes in μ_k from that at room temperature, except for 20 kGy.

The deterioration of frictional properties of the physically cross-linked gels at 37 °C is possibly caused by a decrease in shear strength and an increase in adhesion, because the gel slightly dissolves due to the breakage of physical cross-linking near body temperature. In the gels irradiated with gamma rays, the physical cross-linking destruction was suppressed by the formation of chemically cross-linking points. The above results revealed that the gels irradiated using gamma rays of 10 kGy at 5 kGy/h exhibited better friction characteristics than the conventional (non-irradiated) CD on FT hybrid gels in the reciprocating frictional test.

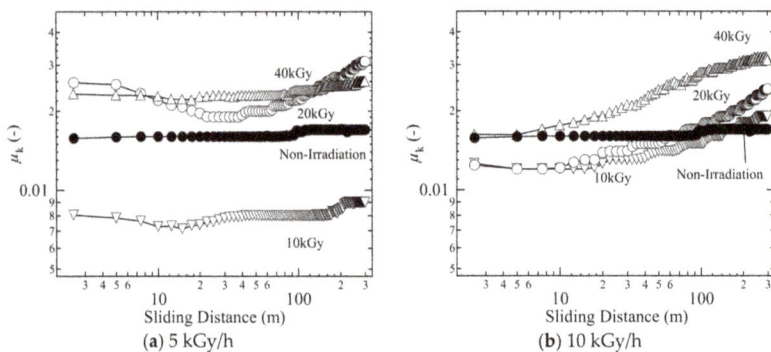

Figure 3. Coefficient of friction of gamma ray irradiated gels with (**a**) 5 kGy/h and (**b**) 10 kGy/h in reciprocating friction tests at body temperature (about 37 °C); close circle, non-irradiation gel; inverted triangle, irradiance = 10 kGy; open circle, irradiance = 20 kGy; triangle, irradiance = 40 kGy.

2.3. Wear Properties

As far as application as artificial cartilage is concerned, wear is as important as friction. Wear should be kept to the minimum in order to suppress adverse effects in the body. Therefore, the sample surface after the reciprocating friction test was observed using a phase contrast microscope. Figure 4 shows a phase contrast micrograph of wear traces. Figure 4a is a photomicrograph of the gel surface after the reciprocating friction test at room temperature (Figure 2). On the surface of gels irradiated at 10 kGy, little scratches are detected. By increasing gamma ray irradiance, scratches caused by wear increased after the reciprocating friction test.

Figure 4b is a photomicrograph of the gel surface after the reciprocating friction test at body temperature (Figure 3). The distinct scratches were formed on the surface of the non-irradiated gel surface at 37 °C. In contrast, the irradiated surface features revealed that wear was suppressed by gamma ray irradiation.

The scratches on the irradiated gel surface except for 10 kGy after rubbing at room temperature are regarded to be caused by the excessive increase in the surface hardness via the formation of chemical cross-linking points. Wear suppression at 37 °C by gamma ray irradiation is considered to be actualized by the improvement in gel structure. The chemical cross-linking point was not destroyed at 37 °C, and the decrease in the shear strength of the surface was suppressed.

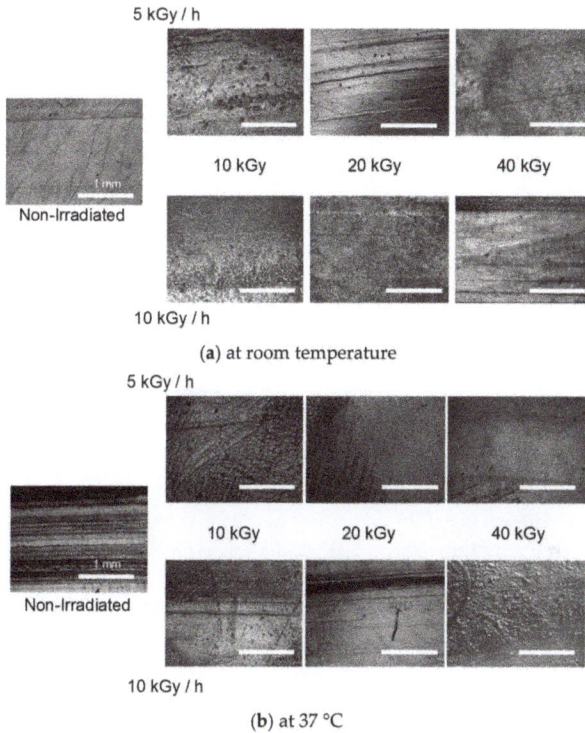

Figure 4. The photomicrograph of the surface of the gel after reciprocating friction tests at (a) room temperature and (b) body temperature. Scale bar is 1 mm.

2.4. Discussion on Friction and Wear Properties

As discussed above, low dose rate and low irradiance appear to actualize good friction and wear characteristics for clinical applications at body temperature.

For non-irradiated gel at 37 °C, the increase in friction and surface scratching is considered to have been brought about by a decrease in shear strength and an increase in adhesive force, as a result of the local breakage of physical cross-linking accompanied with some dissolution of gel. In order to maintain low friction and minimal wear at 37 °C, appropriate improvement of gel structure and properties is required but excessive hardening with cross-linking may deteriorate the tribological properties of PVA CD on FT hybrid gel. The irradiation treatment by gamma ray at 10 kGy and 5 kGy/h is expected to be an optimum condition for gel structure, although further investigation is required to elucidate the detailed improving mechanism with gamma irradiation.

3. Conclusions

Because the chemically cross-linking point was introduced by gamma ray irradiation, the water content decreased, which was still maintained about 70%.

Gamma ray irradiated gels at appropriate conditions showed superior friction and wear characteristics near body temperature. Gels at low dose rate were excellent in terms of their friction and wear characteristics. As the irradiance increased, the friction coefficient increased.

It is concluded that sterilization treatment with low dose rate and irradiance is suitable for applying PVA gels as candidate materials for artificial cartilage. However, it is necessary to confirm that the sterilization is completed under this condition.

4. Materials and Methods

4.1. Sample Preparation

4.1.1. PVA CD on FT Hybrid Gel

PVA pre-gel solution (15 wt %) was prepared by dissolving PVA powder (PVA117; Kuraray Co., Ltd., Tokyo, Japan, degree of polymerization: 1700, degree of hydrolysis: 98 to 99 mol %) in pure water at a temperature around 90 °C over a period of 2 h. The PVA pre-gel solution (15.0 g) was then decanted into a polystyrene dish (with an inner diameter of 85 mm). The pre-gel solution was frozen at −20 °C for 8 h and then thawed at 4 °C for 9 h. This process was repeated 4 times. After the freezing and thawing process, the second CD gel layer was prepared. The PVA pre-gel solution (15.0 g) was decanted on the FT gel layer and dried at 8 °C and 50%RH for 7 days in a temperature and humidity controlled chamber (SU-242, ESPEC, Osaka, Japan), before it was treated at 20 °C and 40%RH until the weight became almost constant. The obtained CD on FT hybrid PVA hydrogel film was soaked in 1 L pure water for 48 h.

4.1.2. Gamma Ray-Irradiation

After swelling the fabricated gels in pure water over 24 h, the samples in water were irradiated by gamma rays from a 60Co radiation source. The irradiation experiments were carried out at National Institutes for Quantum and Radiological Science and Technology (Watanuki 1233, Takasaki, Gunma, Japan). Total dose irradiances were 10, 20, and 40 kGy. The dose rates were about 5 and 10 kGy/h.

4.2. Water Contents

To analyze the swelling property, square samples (10 mm × 10 mm) were cut out from the swollen gels. The weight of the swollen sample, W_t, was measured. After the measurement, the gels were dried at 60 °C for 24 h in a temperature-controlled chamber, and the weight of dried sample, W_d, was measured. From these measurements, water content was calculated as $1 - (W_d/W_t)$.

4.3. Reciprocating Friction Test

For each swollen gel, a ball-on-plate reciprocating friction test was conducted by using a friction tester (TriboGear TYPE:38, HEIDON). Rectangle samples (40 mm × 10 mm) were cut out from the swollen gel. The lubricant temperatures were room temperature and 37 °C. Sliding speed was 20 mm/s. Total sliding distance was 300 m (stroke: 25 mm; total cycle: 6000). Vertical load was 5.88 N. The lubricant was pure water. These conditions were selected according to a previous study [13] to evaluate friction and wear properties of PVA hydrogels in mixed or boundary lubrication mode. Polycrystalline alumina ceramic ball of 26 mm diameter (surface roughness $R_a < 0.01$ μm) was used as an upper ball specimen. From the measured tangential force, coefficient of friction, μ_k, was calculated.

Acknowledgments: This work was supported by JSPS KAKENHI Grant Number JP23000011. And the authors would like to thank Kuraray Co., Ltd. for kindly supplying PVA powders. The authors would like to thank Enago (www.enago.jp) for the English language review.

Author Contributions: Saori Sasaki and Teruo Murakami conceived and designed the experiments; Saori Sasaki performed the experiments and analyzed the data; Seiji Omata contributed sample preparation tools; Naotsugu Nagasawa and Mitsumasa Taguchi performed gamma ray irradiation; Atsushi Suzuki optimized the method for preparing PVA CD on FT hybrid gel; Saori Sasaki wrote the paper.

Conflicts of Interest: The authors declare no conflict of interest.

References

1. Tamura, K.; Ike, O.; Hitomi, S.; Isobe, J.; Shimizu, Y.; Nambu, M. A new hydrogel and its medical application. *ASAIO J.* **1986**, *32*, 605–608. [CrossRef]

2. Otsuka, E.; Suzuki, A. A simple method to obtain a swollen PVA gel crosslinked by hydrogen bonds. *J. Appl. Polym. Sci.* **2009**, *114*, 10–16. [CrossRef]

3. Otsuka, E.; Suzuki, A. Swelling properties of physically cross-linked PVA gels prepared by a cast-drying method. In *Gels: Structures, Properties, and Functions*; Springer: Berlin/Heidelberg, Germany, 2009; pp. 121–126, ISBN 978-3-642-00864-1.

4. Suzuki, A.; Sasaki, S. Swelling and mechanical properties of physically crosslinked poly(vinyl alcohol) hydrogels. *Proc. Inst. Mech. Eng. Part J J. Eng. Med.* **2015**, *229*, 828–844. [CrossRef] [PubMed]

5. Mabuchi, K.; Tsukamoto, Y.; Yamamoto, M.; Ueno, M.; Sasada, T.; Nambu, M. Lubrication property and mechanical durability of the joint prostheses containing cartilage-like polyvinyl alcohol gel. *J. Soc. Orthop. Biomech.* **1986**, *8*, 101–105.

6. Sasada, T. Biomechanics and biomaterials–friction behaviour of an artificial articular cartilage. In *Transactions of the 3rd World Biomaterials Congress*; Kyoto International Conference Hall: Kyoto, Japan, 1988.

7. Murakami, T.; Sawae, Y.; Higaki, H.; Ohtsuki, N.; Moriyama, S. The adaptive multimode lubrication in knee prostheses with artificial cartilage during walking. *Tribol. Ser.* **1997**, *32*, 371–382.

8. Murakami, T.; Yarimitsu, S.; Nakashima, K.; Sawae, Y.; Sakai, N.; Araki, T.; Suzuki, A. Time-dependent frictional behaviors in hydrogel artificial cartilage materials. In Proceedings of the 6th International Biotribology Forum, Fukuoka, Japan, 5 November 2011.

9. Murakami, T.; Yarimitsu, S.; Nakashima, K.; Yamaguchi, T.; Sawae, Y.; Sakai, N.; Suzuki, A. Superior lubricity in articular cartilage and artificial hydrogel cartilage. *Proc. Inst. Mech. Eng. Part J J. Eng. Tribol.* **2014**, *228*, 1099–1111. [CrossRef]

10. Suzuki, A.; Sasaki, S.; Sasaki, S.; Noh, T.; Nakashima, K.; Yarimitsu, S.; Murakami, T. Elution and wear of PVA hydrogels by reciprocating friction. In Proceedings of the 5th World Tribology Congress, WTC 2013, Politecnico di Torino (DIMEAS), Torino, Italy, 8–13 September 2013.

11. Murakami, T.; Sakai, N.; Yamaguchi, T.; Yarimitsu, S.; Nakashima, K.; Sawae, Y.; Suzuki, A. Evaluation of a superior lubrication mechanism with biphasic hydrogels for artificial cartilage. *Tribol. Int.* **2015**, *89*, 19–26. [CrossRef]

12. Murakami, T.; Yarimitsu, S.; Nakashima, K.; Sakai, N.; Yamaguchi, T.; Sawae, Y.; Suzuki, A. Synergistic Lubricating Function with Different Modes for Artificial Hydrogel Cartilage. In Proceedings of the 8th International Biotribology Forum and the 36th Biotribology Symposium, Yokohama Japan, 21–22 September 2015.

13. Sasaki, S.; Murakami, T.; Suzuki, A. Frictional properties of physically cross-linked PVA hydrogels as artificial cartilage. *Biosurf. Biotribol.* **2016**, *2*, 11–17. [CrossRef]

14. Tohfafarosh, M.; Baykal, D.; Kiel, J.W.; Mansmann, K.; Kurtz, S.M. Effects of gamma and e-beam sterilization on the chemical, mechanical and tribological properties of a novel hydrogel. *J. Mech. Behav. Biomed. Mater.* **2016**, *53*, 250–256. [CrossRef] [PubMed]

15. Kanjickal, D.; Lopina, S.; Evancho-Chapman, M.M.; Schmidt, S.; Donovan, D. Effects of sterilization on poly (ethylene glycol) hydrogels. *J. Biomed. Mater. Res. Part A* **2008**, *87*, 608–617. [CrossRef] [PubMed]

16. Wang, B.; Kodama, M.; Mukataka, S.; Kokufuta, E. On the intermolecular crosslinking of PVA chains in an aqueous solution by γ-ray irradiation. *Polym. Gels Netw.* **1998**, *6*, 71–81. [CrossRef]

17. Rosiak, J.M.; Yoshii, F. Hydrogels and their medical applications. *Nucl. Instrum. Methods Phys. Res. B* **1999**, *151*, 56–64. [CrossRef]

18. Bhat, N.V.; Nate, M.M.; Kurup, M.B.; Bambole, V.A.; Sabharwal, S. Effect of γ-radiation on the structure and morphology of polyvinyl alcohol films. *Nucl. Instrum. Methods Phys. Res. B* **2005**, *237*, 585–592. [CrossRef]

19. Schulz, R.M.; Bader, A. Cartilage tissue engineering and bioreactor systems for the cultivation and stimulation of chondrocytes. *Eur. Biophys. J.* **2007**, *36*, 539–568. [CrossRef] [PubMed]

gels

MDPI

Article

Frozen State of Sephadex® Gels of Different Crosslink Density Analyzed by X-ray Computed Tomography and X-ray Diffraction

Norio Murase [1,*,†], Yuki Uetake [1], Yuki Sato [1], Kentaro Irie [2], Yohei Ueno [2], Toru Hirauchi [2], Toshio Kawahara [2] and Mitsuhiro Hirai [3]

1   Division of Life Science and Engineering, School of Science and Engineering, Tokyo Denki University, Hatoyama, Hiki-gun, Saitama 350-0394, Japan; uetake.yuki0414@gmail.com (Y.U.); yuuki_satou01@yahoo.co.jp (Y.S.)
2   Nisshin Seifun Group Inc., Research Center for Basic Science, Research and Development, Quality Assurance Division, 5-3-1 Tsurugaoka, Fujimino-city, Saitama 356-8511, Japan; irie.kentaro@nisshin.com (K.I.); ueno.yohei@nisshin.com (Y.U.); hirauchi.toru@nisshin.com (T.H.); kawahara.toshio@nisshin.com(T.K.)
3   Graduate School of Science and Technology, Gunma University, 4-2 Aramaki, Maebashi, Gunma 371-8510, Japan; mhirai@gunma-u.ac.jp
*   Correspondence: nmurase@mail.dendai.ac.jp; Tel.: +81-48-824-6588
†   Present Affiliation: Professor Emeritus of Tokyo Denki University.

Received: 12 March 2018; Accepted: 6 May 2018; Published: 18 May 2018

Abstract: Water in Sephadex® (crosslinked dextran) gels is known to indicate different freezing behavior which is dependent on the density of the crosslinks, and water in a Sephadex® G25 gel remains partially unfrozen during cooling and crystallizes during rewarming. The mechanism of anomalous ice crystallization during rewarming is still unclear. The objective of this study is to observe the ice grains that form in Sephadex® beads and to comprehend their frozen state with a focus on the ice crystallization during rewarming. Sephadex® beads containing 50 wt % water were prepared and used for the measurements. The observation of the ice grains was carried out by using synchrotron radiation-sourced X-ray CT (computed tomography). XRD (X-ray diffraction) analysis was also conducted to investigate the frozen state. As a result, ice grains that were larger than ~1 μm were hardly observed after the slow cooling of Sephadex® beads, except in the G25 beads. However, at the occurrence of ice crystallization during rewarming, ice grains that were larger than 10 μm appeared in the G25 beads. Using XRD, it was found that small incomplete ice crystals were formed in G25 beads and the presence of glassy water was indicated in the gel. In conclusion, the size and distribution of ice grains that formed in Sephadex® beads were different depending on the density of the crosslinks.

Keywords: Sephadex® (crosslinked dextran); crosslink density (density of crosslinks); ice grain; ice crystallization during rewarming; glassy water; X-ray CT; XRD

---

1. Introduction

Biological systems often take a gelled state. Understanding the frozen state of gels (i.e., the size and distribution of ice grains) is of practical importance for the implementation of cryopreservation. Water in gels is compartmentalized by the polymer network [1,2], and there is a possibility that the polymer network obstructs and retards the diffusional motion of the water molecules in gels [3], consequently preventing the growth of ice crystals. The study of the freezing behavior and the state of the frozen gels is of basic interest, as they reflect the characteristics of the polymer network (i.e., the flexibility of polymer chains, the size of the compartments, and the extent of the continuity between adjacent compartments which are interrelated via the density of the crosslinks) [4].

In this connection, the freezing behavior of Sephadex® (crosslinked dextran) gels has been investigated mainly by DSC (differential scanning calorimetry) [1,2] and by various physical techniques [5]. Stemming from these investigations, different freezing behaviors were indicated depending on the density of the crosslinks. It was found that with the Sephadex® G25 gel, the water remains partially unfrozen during cooling and the ice crystallizes during the subsequent warming (rewarming), as indicated by DSC. A conceivable scheme of ice crystallization during rewarming is described as follows. Some part of the water in the gel is trapped by the polymer network at the time of freezing initiation when it is cooled, followed by a change in the polymer network that turns it into a glassy state. The glassy water crystallizes during the rewarming, caused by the partial melting of the ice that was previously separated [5–7]. The presence of glassy water, however, remains unconfirmed. Even when the water does not turn into a glassy state, small ice crystals are formed in the gel and a powder diffraction pattern has been observed by a two-dimensional X-ray diffraction (XRD) study [8]. In Sephadex® gels with crosslinks of both higher and lower densities than that of G25, an anomalous freezing behavior during rewarming was not observed, and the larger ice crystals were considered to have formed, which was also indicated by the two-dimensional XRD study [8].

As the glass transition temperature (T_g) of hyper-quenched liquid water is known to be around 136 K (-137 °C) [9], the T_g of water in a G25 gel (assumed to be around -50 °C) is too high. The T_g of the water in the G25 gel was estimated from the temperature where the liquid water disappears in the ESR (Electron Spin Resonance) spectra obtained by using a spin probe method. The water in the G25 gel is partially trapped by the polymer network and turns into a glassy state at the time of freezing initiation, presumably as a consequence of the close interaction with the hydroxy groups located at the surface of the dextran network structures via the hydrogen bonds. The high T_g value of the water in the gel can thus be explained. In fact, the T_g of bulk water confined in MCM-41 nanopores was suggested to be 210 K [10].

The occurrence of ice crystallization during rewarming observed with Sephadex® G25 gel is characteristic of non-equilibrium freezing after substantial supercooling, and it is not observed after equilibrium freezing [7]. In this connection, supercooled water is of great interest from a scientific standpoint and has been actively discussed since the 1970s [11]. Although various thermodynamic properties such as the heat capacity (C_p), the isothermal compressibility, and so forth have indicated a divergence at around -45 °C, the behavior of said divergence is a matter of concern [11,12] and it is still unclear due to the interference by homogeneous nucleation which occurs at around -40 °C. Concerning the amorphous phases of ice, the high-density and low-density amorphous forms of ice and/or polymorphism—where a deeper understanding of the hydrogen bond network in water is required [13]—have also been intensively discussed. Although the presence of supercooling is a requisite for the observation of ice crystallization during rewarming, the water in the gel freezes at around -22 °C; the degree of supercooling is not very high. Then, the rates of freezing and the resultant change of the polymer network dependent on the degree of supercooling need to be fast for the observation and the occurrence of the glass transition. The analysis of ice crystallization during rewarming observed with a G25 gel might provide valuable information in this connection.

Ice crystals exist in aggregates as ice grains, and small ice grains are formed as a consequence of the formation of small ice crystals. Then, the sizes of the ice grains observed reflect the sizes of ice crystals that are formed. Although it might be true that the size of ice crystals (and therefore the ice grains formed) are different depending on the density of the crosslinks or on the polymer network structure, the information obtained by the diffraction pattern of the two-dimensional XRD is unsatisfactory, as there is no information on the distribution of the ice grains in the gels. The direct observation of ice grains at a high resolution is desirable for the understanding of the frozen state. The observation of the beads by SEM (scanning electron microscope) after freeze-drying is possible.

However, there is concern that the technique will lead to changes in the polymer network during the freeze-drying treatment.

In the present study, the in situ observation of the ice grains formed in Sephadex® gel beads was conducted by the X-ray CT (computed tomography) imaging technique. Recently, the X-ray CT technique has been improved, making it possible to acquire three-dimensional images of the microstructure of foods or biological systems [14,15]. The technique of X-ray CT imaging is based on the detection of the differences in the X-ray absorption rate by the materials that compose the sample. The contrast of the images depends on the wavelength, chromaticity (mono- or poly-), and the brilliance of the X-ray beam applied, as well as the mass density and chemical components of materials. By using synchrotron radiation, a highly brilliant and monochromatic X-ray beam can be provided and a high spatial resolution was made possible. Then, the synchrotron radiation-sourced X-ray CT analysis of the ice grains formed in Sephadex® beads was conducted at SPring-8, a synchrotron radiation facility in Japan. The detection of ice grains in the micrometer range by using a beam line in the facility was anticipated. The observation of the freeze-dried Sephadex® beads by SEM was also conducted for comparison.

To understand the frozen state of gels, it is necessary to obtain information about the ordered structure of the polymer network forming the gel as well as the size of the ice grains that are smaller than a micrometer. For that purpose, synchrotron radiation-sourced SAXS (small-angle X-ray scattering) measurements were also carried out at SPring-8. For the analysis of the completeness of ice crystals determining the shape of the ice grains, WAXD (wide-angle X-ray diffraction) measurements were carried out together with SAXS measurements in the XRD study. As the XRD study deals with the structure of dimensions that are smaller than the beads, the term "gel" was used for the XRD experiment instead of "beads."

The objective of the present study was to analyze the shape and distribution of the ice grains formed in the Sephadex® beads with different crosslink densities, as well as to comprehend the frozen state, especially that of the G25 gel, that leads to ice crystallization during rewarming.

2. Results

2.1. X-ray CT Measurement

Typical X-ray CT images obtained with Sephadex® beads after rapid cooling are shown in Figure 1. Linear X-ray absorption coefficients (LACs) obtained for ice and the bead matrix in the figure were about 2.2 and 3.0, respectively. The LAC order was quartz capillary > bead matrix > ice > air, and the bead matrix observed in the capillary is represented as gray circles. The ice grains are represented in a gray color slightly darker than the bead matrix.

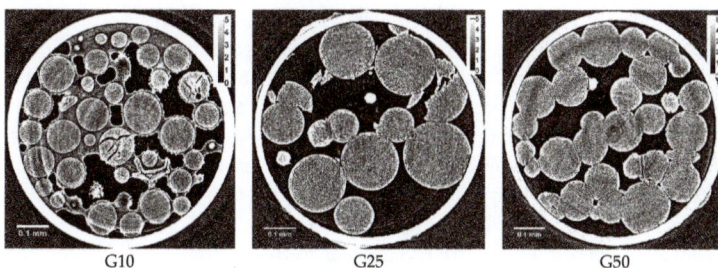

G10　　　　　　　　G25　　　　　　　　G50

Figure 1. X-ray computed tomography (CT) images obtained with rapidly cooled Sephadex® beads containing 50 wt % water: (**left**) G10; (**middle**) G25; (**right**) G50 beads. Small white circles different from the beads observed in G25 and G50 beads might be artifacts during the measurement. Calibration bars on the upper right indicate the linear X-ray absorption coefficient (LAC, cm^{-1}). The density of the crosslinks: G10 > G15 > G25 > G50 > G100 [16].

In the case of G10 beads, ice due to unabsorbed water was seen around the beads. Slits observed dark in some beads were due to the cracks formed during freezing. However, spots due to the ice grains were not identified in the internal region of G10 beads, indicating that ice grains smaller than ~1 μm were formed in the beads. In the internal region of G25 and G50 beads, the ice grains were not identified either. Several fragments of the beads observed in the image of G25 beads were probably made by breakage during the sample packing into the capillary.

Typical X-ray CT images obtained with the beads after slow cooling are shown in Figure 2.

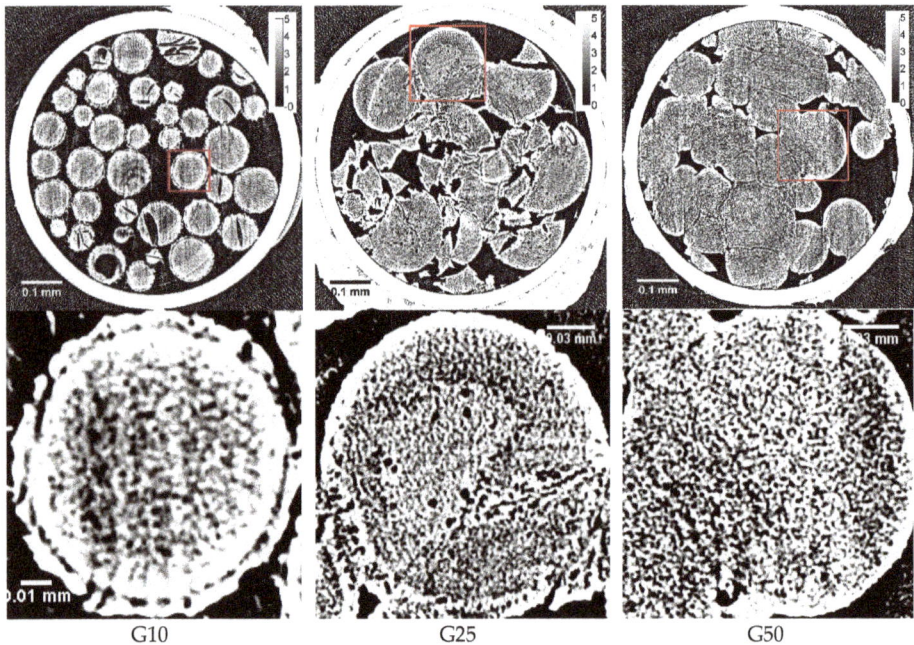

Figure 2. X-ray CT images obtained with slowly cooled Sephadex® beads containing 50 wt % water (**top**) together with the expanded images of the square part of the corresponding upper ones (**bottom**): (**left**) G10; (**middle**) G25; (**right**) G50. Calibration bars on the upper right indicate the LAC (cm^{-1}).

Among the three kinds of Sephadex® beads, there was a difference in the appearance of the surface as well as in the degree of the connection between adjacent beads. G10 beads were isolated from each other. On the other hand, adjacent G50 beads were ready to stick together. The degree of stickiness of G25 beads resided between G10 and G50 beads. Large spots observed on the margin of G10 beads were probably due to the ice grains. With G25 beads, large spots were also observed on the margin of the beads. However, no large ice grains were observed on the margin of G50 beads.

The dark spots of 3-5 μm in diameter observed in the internal region of G25 beads were due to the ice grains, though they are not very abundant. With G10 and G50 beads, spots due to the ice grains larger than ~1 μm were not identified in the internal region.

X-ray CT images obtained with refrozen G25 beads re-cooled from −9 °C (the temperature of the completion of the ice crystallization exotherm) are shown in Figure 3 together with the DSC heating trace. Large ice grains larger than 10 μm appeared after the occurrence of ice crystallization during rewarming. The dependence of the X-ray CT images of refrozen G25 beads on the temperature from which re-cooling was initiated is shown in Figure 4. When the re-cooling was initiated from −9 °C, large ice grains observed in the X-ray CT image were most abundant among the images different in the initiation temperature of re-cooling, −11 or −5 °C.

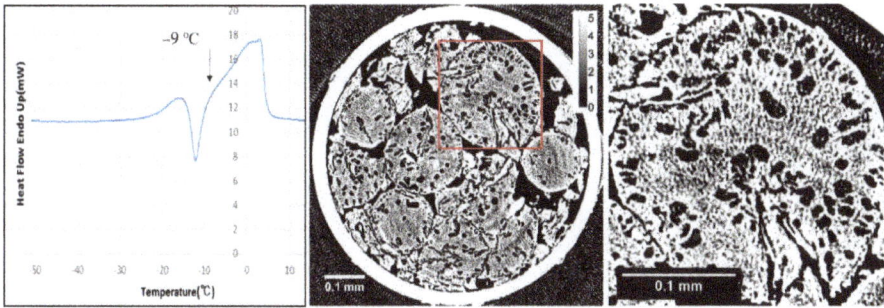

Figure 3. X-ray CT images obtained with refrozen Sephadex® G25 beads re-cooled from −9 °C (**middle**) together with the expanded image of the square part (**right**), and a differential scanning calorimetry (DSC) rewarming trace indicating the ice crystallization exotherm (**left**). DSC cooling and heating rates were 5 °C min^{-1}.

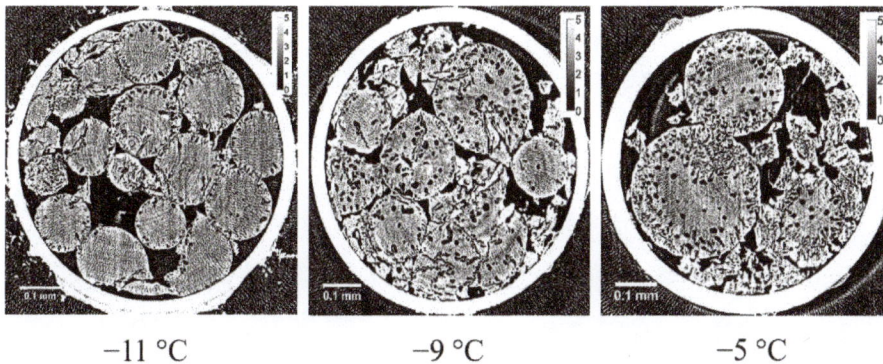

$$-11\,°C \qquad -9\,°C \qquad -5\,°C$$

Figure 4. Dependence of X-ray CT images of refrozen G25 beads on the temperature from which the re-cooling was initiated: (**left**) −11 °C; (**middle**) −9 °C; (**right**) −5 °C.

2.2. Observation by SEM

Sephadex® beads observed by SEM after freeze-drying are shown in Figure 5. In the G10 bead, a crack was clearly observed. The surface of the bead appeared to be smooth and rigid. With the G25 bead, hollows indicating the trace of the ice grains were observed in the internal region of the beads as well as on the margin. These characteristics of the SEM images correspond to those obtained by X-ray CT. However, clear difference in the network structure of the bead matrix was not confirmed among the Sephadex® beads used.

2.3. XRD Measurement

The SAXS measurement was carried out to obtain information about the structure of the polymer network or the size of the ice grains smaller than the micrometer order. However, characteristics dependent on the density of crosslinks were not confirmed with the frozen Sephadex® gels. On the other hand, in the WAXD measurement, Bragg peaks due to the hexagonal ice observed at around 0.155-0.175 for the scattering vector q (10 × nm^{-1}) indicated the different temperature dependence among the three Sephadex® gels, G15, G25, and G100, as shown in Figure 6.

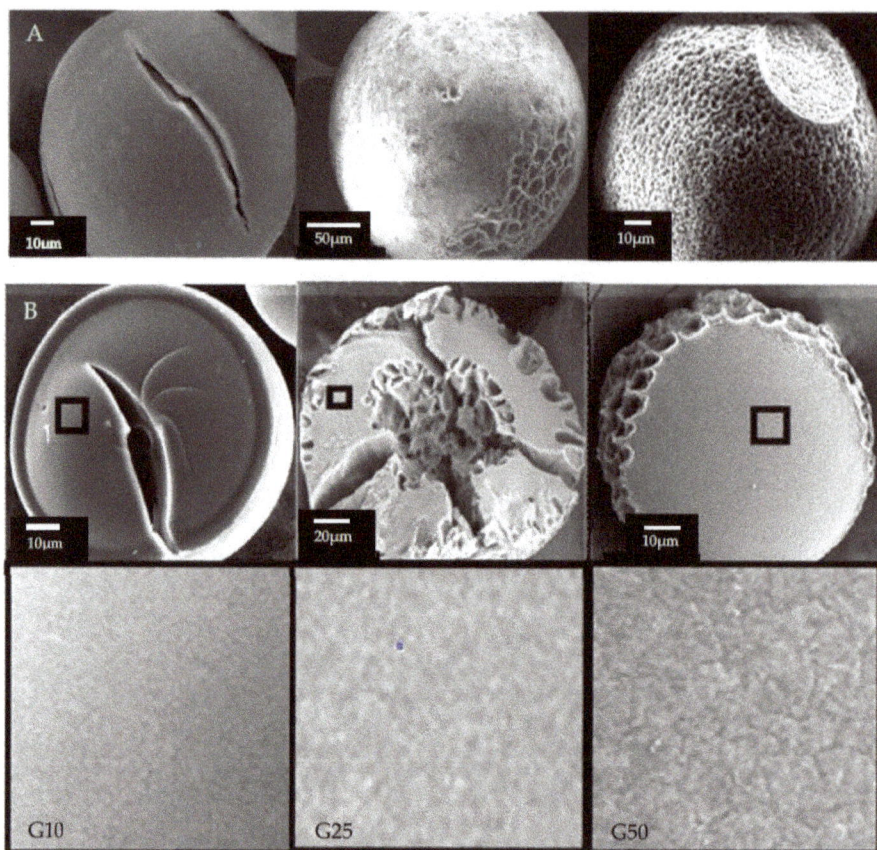

Figure 5. SEM images of Sephadex® beads after freeze-drying. (**A**) Beads; (**B**) Surface of the fractured beads (**top**) and the expanded images of the square part of the corresponding upper ones (**bottom**): (**left**) G10; (**middle**) G25; (**right**) G50.

Bragg peaks of a frozen G25 gel observed at −40 °C showed a remarkable tailing suggesting the formation of small incomplete ice crystals or the presence of glassy water. With the increase in temperature, the peaks became narrow at temperatures between −20 and −15 °C, where an ice crystallization exotherm during rewarming was initiated in the DSC trace. The narrowing trend continued to −5 °C and the peaks disappeared at temperatures between −5 and 0 °C, where ice melted. Bragg peaks observed with a frozen G25 gel depending on the freezing condition are shown in Figure 7. When the heating of the frozen G25 gel was interrupted at the temperature of the completion of ice crystallization during rewarming followed by re-cooling, Bragg peaks of the refrozen G25 gel became narrow without tailing through the temperature from −40 °C to the melting temperature, suggesting the formation of large stable ice crystals at the occurrence of crystallization.

Bragg peaks of a frozen G15 gel observed at −40 °C also showed a slight tailing (Figure 6), which weakened with the increase in temperature. On the other hand, in the case of a frozen G100 gel, narrow Bragg peaks were observed through temperatures ranging from −40 °C to the melting temperature, indicating the formation of large stable ice crystals during cooling.

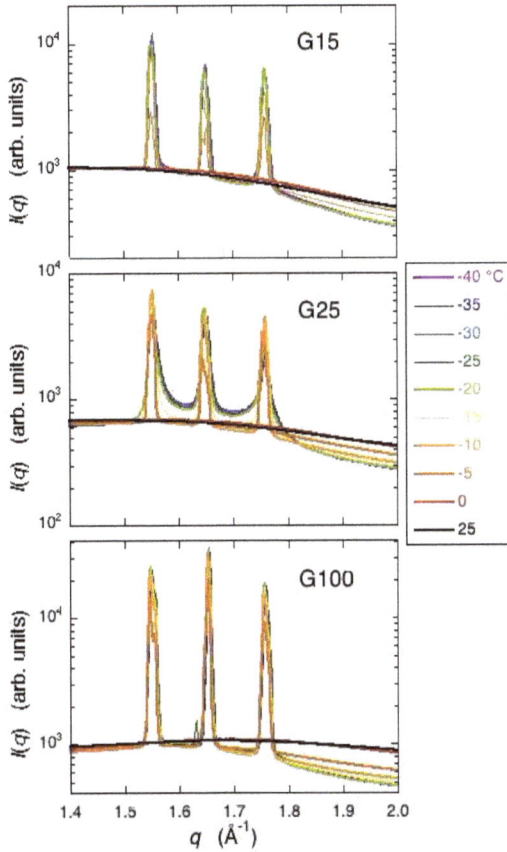

Figure 6. Temperature dependence of the WAXD (wide-angle X-ray diffraction) curves obtained with frozen Sephadex® gels. G15, G25, and G100 gels from the top to the bottom, respectively.

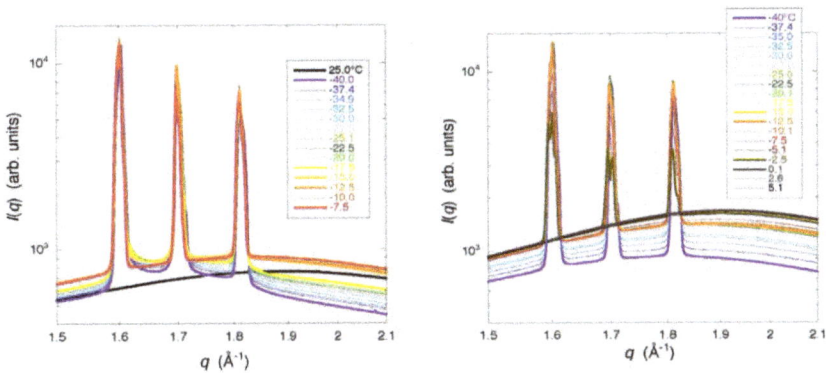

Figure 7. Temperature dependence of the WAXD curves obtained with frozen and refrozen G25 gels. (**left**) Bragg peaks observed with a frozen G25 gel; (**right**) Bragg peaks observed with a refrozen G25 gel re-cooled from −7.5 °C. The temperature interval in this experiment was 2.5 °C.

3. Discussion

By the present study of X-ray CT measurement using synchrotron radiation, the in situ observation of the ice grains in Sephadex® beads was made possible with a spatial resolution of ~1 μm. As a result, it is necessary to change the interpretation of the size and distribution of the ice grains formed in the beads that was held previously.

By the previous study of the two-dimensional XRD measurement, it was interpreted that ice crystals of larger size were formed in a G10 gel compared to those in G25 and G100 gels, as many clear spots were observed on the three concentric rings due to the diffractions by hexagonal ice in the image [8]. Moreover, it was considered that the formation of large ice crystals corresponded to the formation of the large ice grains in the internal region of G10 beads. However, clear spots observed by the two-dimensional XRD measurement were probably due to the ice crystallized outside the G10 beads as the unabsorbed water existed, which was indicated by the X-ray CT image obtained after rapid cooling (Figure 1). Although the water holding capacity of a G10 gel is 1 g water/g dry gel, corresponding to a water content of 50 wt % according to the literature [16], it is reasonable to consider that some water remains unabsorbed outside the gel beads because of the inhomogeneous distribution of water in the bead sample. There is another possibility of the observation of clear spots in the two-dimensional XRD image of a G10 gel—the large ice grains may be formed in the region of the beads between the rigid surface-treated layer and the polymer matrix. The existence of this region is suggested by the image in Figure 2. However, no ice grains larger than ~1 μm were observed in the internal region of G10 beads, even after slow cooling, though most of the water in the beads was frozen. By the DSC measurement, about 70 percent of the water in the Sephadex® beads was estimated to be frozen during slow cooling [2]. Looking at the expanded X-ray CT images of the beads (Figure 2), small structures of ~1 μm seemed to appear in the internal region of the beads. However, they might be artifacts, as the fluctuation of the X-ray absorption rate is considerable, especially at the border between ice and the polymer matrix. Ring artifacts were also observed, as can especially be seen in Figure 1.

In the internal region of G25 beads, a small number of ice grains of 3-5 μm in diameter were observed after slow cooling. Although the ice grains were observed in G25 beads, most of the ice grains that formed in the rest of the beads were smaller than ~1 μm, as they were not visible in the image. This is consistent with the results obtained from the two-dimensional XRD study where the powder diffraction pattern was observed [8]. In the X-ray CT image obtained with the refrozen G25 beads re-cooled from −9 °C (the temperature of the completion of ice crystallization during rewarming), many large ice grains larger than 10 μm appeared in the internal region of the beads, which is also consistent with the result of the two-dimensional XRD study where the spotty diffraction pattern indicating the formation of large ice crystals was observed.

There were no surface-treated layers observed in G50 beads. Instead, G50 beads containing 50 wt % water were ready to stick together, probably by means of the tangling bonds of the polymer chains existing on the bead surface. In this connection, G100 beads with lower crosslink densities than G50 beads became too sticky, losing the bead structure when they were added with water [17], making it difficult to prepare G100 bead samples in quartz capillaries. That is the reason why G50 beads were used in this study instead of G100 beads as the bead sample with low crosslink densities. No ice grains were observed in the CT image of G50 beads. Considering the result obtained by the two-dimensional XRD study using a G100 gel, where pairs of arcs facing each other were seen on the three concentric rings due to the diffractions by hexagonal ice [8], there is a possibility that long and narrow structures could be observed in the X-ray CT image of G50 beads. However, the width of the structures may have been too narrow (<~1 μm), if they existed, to be recognized by the observation.

From the results of the WAXD measurement, the formation of small incomplete ice crystals in a G25 gel was indicated, as Bragg peaks showed a remarkable tailing. The result is consistent with that of the two-dimensional XRD study where continuous but dim images due to the diffractions by powder crystals was obtained for hexagonal ice. Moreover, the presence of glassy water in a frozen

G25 gel was indicated, though not decisive [5,18]. A slight tailing in the Bragg peaks was also observed with a frozen G15 gel, indicating the presence of small ice crystals in the gel with higher crosslink densities than a G25 gel. The finding is consistent with the interpretation of the result obtained with a G10 gel by the X-ray CT measurement. In the case of a G100 gel with lower crosslink densities than a G25 gel, the formation of large and stable ice crystals was indicated by WAXD, which is also consistent with the result obtained by the two-dimensional X-ray measurement, though the presence of long and narrow ice structures was not verified by the X-ray CT measurement as mentioned above. The observation of narrow Bragg peaks with a refrozen G25 gel indicated large and stable ice crystals. Therefore, large ice grains were formed in the beads at the time of ice crystallization during rewarming. This result corresponds with that of the X-ray CT measurement.

It is certain that the frozen state of the Sephadex® gels depending on the density of the crosslinks was made clearer by the experiments of a synchrotron radiation-sourced X-ray CT and an XRD. However, the frozen state of the gels of submicron order remains unclarified. More precise SAXS measurements are necessary. Concerning ice crystallization during rewarming, the presence of glassy water in a G25 gel is still unconfirmed. Detection of the glass transition in the frozen gel is desirable for this clarification [10].

4. Conclusions

The frozen state of Sephadex® gels containing 50 wt % water was analyzed by using a synchrotron radiation-sourced X-ray CT and XRD.

By X-ray CT, the ice grains in Sephadex® beads were observed with a spatial resolution of ~1 μm. As a result, ice grains larger than ~1 μm were hardly observed in the beads after slow cooling, independent of the crosslink density. In G25 beads, a small number of ice grains of 3-5 μm in diameter were observed after slow cooling, but most of the ice grains formed in the beads were smaller than ~1 μm. Ice grains larger than 10 μm in diameter were successfully observed in the refrozen G25 beads re-cooled from the temperature of the completion of ice crystallization during rewarming. These results obtained by the X-ray CT measurement were consistent with those previously obtained by two-dimensional XRD measurement.

The results obtained by XRD—especially by WAXD—corresponded well with those obtained by X-ray CT. The observation of a remarkable tailing in Bragg peaks due to the diffraction by hexagonal ice indicated the formation of small incomplete ice crystals or the presence of glassy water in a G25 gel. With the refrozen G25 gel re-cooled from the temperature of the completion of ice crystallization during rewarming, the tailing in Bragg peaks disappeared, which indicated the formation of large and stable ice crystals, as were observed by X-ray CT.

Finally, the size and distribution of the ice grains formed in Sephadex® beads were found to depend on the crosslink density.

5. Materials and Methods

5.1. Materials

Sephadex® beads with different crosslink densities (i.e., G10, G15, G25, G50, and G100) were obtained from GE Healthcare UK Ltd. (Little Chalfont, Buckinghamshire, UK). The size of the beads' diameter was 40–120 μm for G10, G15, and G100 and 50-150 μm for G25 and G50, and the order of the density of the crosslinks is G10 > G15 > G25 > G50 > G100. The exclusion limits in Mw/Da reported are 700, 1500, 5000, 30,000, and 100,000, and the amounts of water absorption in the swollen state (mL g^{-1}$_{DM}$) are 1.0, 1.5. 2.5, 5.0, and 10.0 for G10, G15, G25, G50, and G100, respectively [16]. In the compartment of G25, 250-300 water molecules could be accommodated, assuming a spherical form [2]. The water content of the beads was adjusted to 50 wt % by the addition of distilled water, as the ice crystallization exotherm during rewarming in the DSC trace was most remarkably observed around the water content with a G25 gel.

5.2. X-ray CT Measurement

Sephadex® G10, G25, and G50 were used in this experiment. G100 beads became too sticky, losing their bead structure when they were added with water, making it difficult to prepare G100 bead samples in quartz capillaries. So, G50 beads were used for the X-ray CT measurement instead of G100 beads as the bead sample with low crosslink densities (see the Discussion section). After having been dried at 60 °C for 20 h, each Sephadex® bead sample containing 50 wt % water was prepared by the addition of distilled water to the dried beads. For the preparation of samples used in the X-ray CT measurement, the beads were densely packed by 2-3 mm length in a quartz capillary with an outside diameter of 0.7 mm and about 10 mm length. Both ends of the capillary containing bead samples were then sealed with silicon grease to prevent the evaporation of water. Samples prepared were cooled from 20 to −40 °C at the rate of 1 °C min⁻¹ by using a programmable freezer and kept below −40 °C until the measurement. In the case of G25, the refrozen samples re-cooled from the temperature of the completion of ice crystallization during rewarming (−11 to −5 °C) to −40 °C were also prepared. Cooling and re-cooling rates were 1 °C min⁻¹, and the heating rate during rewarming was 5 °C min⁻¹. Rapid freezing where samples were put into a deep freezer controlled at −85 °C prior to the measurement was also conducted. The cooling rate at that time was ca. 60 °C min⁻¹.

The X-ray CT measurement was carried out using the beam line of BL46XU at SPring-8 (Sayo, Hyogo, Japan). The X-ray wavelength was 0.1 nm, corresponding to 12.4 keV in energy. The angle of X-ray mirrors necessary to eliminate contamination of the X-ray by higher-order harmonics was set at 3.56 mrad. The size of the X-ray beam was 1 mm × 1 mm. The temperature of the frozen samples in the sample holder on the turntable (sample stage) was kept at about −40 °C by regulating the ejection speed of LN$_2$ vapor introduced into the sample holder, and was measured with a thermocouple made of chromel-alumel. Frozen samples set on the turntable were continuously rotated 180 degrees at a rate of 1.2 degrees per second. The transmission images were acquired by using an X-ray imaging system composed of an AA50 X-ray imaging unit and a C4880-41S CCD camera (Hamamatsu Photonics K.K., Hamamatsu City, Shizuoka, Japan) set at the distance of 25 mm from the sample. A photograph of the apparatus for the X-ray CT measurement is shown in Figure 8. A set of 259 transmission images were acquired during the rotation. The exposure time for image acquisition was 250 msec. Two-dimensional images (tomograms) were reconstructed by the filtered back projection method. For each measurement, 1200 pieces of tomograms to be stacked were obtained. The size of pixels in the image data was 0.35 μm × 0.35 μm.

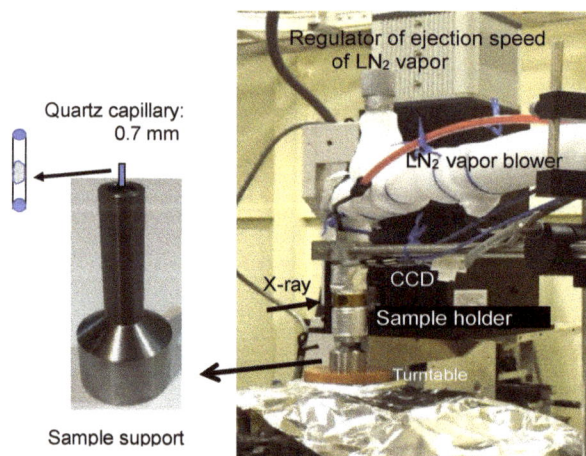

Figure 8. Photograph of an apparatus for X-ray CT measurement.

5.3. Observation by SEM

Samples for the observation by SEM were prepared in the same way as for the X-ray CT measurement. Sample beads of G10, G25, and G50 containing 50 wt % water were cooled to −40 °C at the rate of 1 °C min^{-1} by using a programmable freezer. After being freeze-dried, the beads were fixed on the aluminum stage with carbon tape and were sputter-coated with gold. Freeze-dried beads broken with a hammer were also prepared to observe the inside of the beads. SEM images were taken with a JSM-6010 PLUS_LA (JEOL Ltd., Akishima, Tokyo, Japan).

5.4. XRD Measurement

For the XRD measurement, Sephadex® beads of G15, G25, and G100 were used without drying. Samples containing ca. 50 wt % water were put in cells with 5 mm inner diameter and 2 mm width, and were sealed with a pair of Kapton film windows of 25 μm in thickness. Samples were cooled from 25 to −40 °C at a rate of ca. 13 °C min^{-1}, followed by stepwise heating to 25 °C with a temperature interval of 5 °C for G25 and G100, and 10 °C for G15. Annealing time at each temperature was about 5 min. Measurements of SAXS and WAXD were carried out using the beam line of BL40B2 at SPring-8. The X-ray wavelength was 0.1 nm, and the size of the X-ray beam was 0.1 mm × 0.1 mm. The range for scattering vector q (from the range for SAXS to that for WAXD) covered $0.04 < q \,(10 \times nm^{-1}) < 2.2$, corresponding to 16-0.28 nm of the real space. Camera length was 4153 cm for SAXS and 39 cm for WAXD. Exposure time for data acquisition was 60 s.

Author Contributions: N.M., Y.U., Y.S., K.I., Y.U., T.H. and T.K. conceived, designed, and performed the X-ray CT experiment at SPring-8. They also analyzed the data obtained. N.M. and M.H. conceived and designed, and M.H. performed the XRD experiment at SPring-8. He also analyzed data obtained. N.M. wrote the paper.

Acknowledgments: The authors wish to thank Masugu Sato, Norimichi Sano and Kentaro Kajiwara for their support through the experiment at SPring-8 and for giving a great deal of advice during the X-ray CT measurement. The synchrotron radiation experiments were performed at the BL46XU of SPring-8 with the approval of the Japan Synchrotron Radiation Research Institute (JASRI) Proposal No. 2015A1852, 2015B1924, 2016A1784 and at the BL19B2 of SPring-8 with the approval of the JASRI Proposal No. 2015B1782, 2016A1811 for the X-ray CT measurement, and at the BL40B2 of SPring-8 with the approval of JASRI Proposal No. 2015B1489 for the XRD measurement.

Conflicts of Interest: The authors declare no conflict of interest.

References

1. Murase, N.; Shiraishi, M.; Koga, S.; Gonda, K. Low-temperature calorimetric studies of compartmentalized water in hydrogel systems (I). *Cryo-Letters* **1982**, *3*, 251–254.
2. Murase, N.; Gonda, K.; Watanabe, T. Unfrozen compartmentalized water and its anomalous crystallization during warming. *J. Phys. Chem.* **1986**, *90*, 5420–5426. [CrossRef]
3. Watanabe, T.; Ohtsuka, A.; Murase, N.; Barth, P.; Gersonde, K. NMR studies on water and polymer diffusion in dextran gels. Influence of potassium ions on microstructure formation and gelation mechanism. *Magn. Reson. Med.* **1996**, *35*, 697–705. [CrossRef] [PubMed]
4. Murase, N.; Watanabe, T. Ice crystallization during rewarming of polymer gels. In *Physics and Chemistry of Ice*; Maeno, N., Hondoh, T., Eds.; Hokkaido University Press: Sapporo, Japan, 1992; pp. 249–253, ISBN 4-8329-0261-X.
5. Murase, N.; Ruike, M.; Yoshioka, S.; Katagiri, C.; Takahashi, H. Glass transition and ice crystallization of water in polymer gels, studied by oscillation DSC, XRD-DSC simultaneous measurements, and Raman spectroscopy. In *Amorphous Food and Pharmaceutical Systems*; Levine, H., Ed.; Roy Soc Chem: Cambridge, UK, 2002; pp. 339–346, ISBN 0-85404-866-9.
6. Murase, N. Origin of an endothermic trend observed prior to ice crystallization exotherm in the DSC rewarming trace for polymer gels. *Cryo-Letters* **1993**, *14*, 365–374.
7. Murase, N.; Inoue, T.; Ruike, M. Equilibrium and nonequilibrium freezing of water in crosslinked dextran gels. *Cryo-Letters* **1997**, *18*, 157–164.

8.   Murase, N.; Abe, S.; Takahashi, H.; Katagiri, C.; Kikegawa, T. Two-dimensional diffraction study of ice crystallization in polymer gels. *Cryo-Letters* **2004**, *25*, 227–234. [PubMed]

9.   Johari, G.P.; Hallbrucker, A.; Mayer, E. The glass-liquid transition of hyperquenched water. *Nature* **1987**, *330*, 552–553. [CrossRef]

10.   Oguni, M.; Kanke, Y.; Nagoe, A.; Namba, S. Calorimetric study of water's glass transition in nanoscale confinement, suggesting a value of 210 K for bulk water. *J. Phys. Chem. B* **2011**, *115*, 14023–14029. [CrossRef] [PubMed]

11.   Angell, CA. Supercooled Water. In *Water, A Comprehensive Treatise*; Franks, F., Ed.; Plenum Press: New York, NY, USA, 1982; Volume 7, pp. 1–81, ISBN 0-306-37181-2.

12.   Speedy, R.J.; Angell, C.A. Isothermal compressibility of supercooled water and evidence for a thermodynamic singularity at −45 °C. *J. Chem. Phys.* **1976**, *65*, 851–858. [CrossRef]

13.   Perakis, F.; Amann-Winkel, K.; Lehmkühler, F.; Sprung, M.; Mariedahl, D.; Sellberg, JA.; Pathak, H.; Späh, A.; Cavalca, F.; Schlesinger, D.; et al. Diffusive dynamics during the high-to-low density transition in amorphous ice. *Proc. Natl. Acad. Sci. USA* **2017**, *114*, 8193–8198. [CrossRef] [PubMed]

14.   Kobayashi, R.; Kimizuka, N.; Watanabe, M.; Suzuki, T. The effect of supercooling on ice structure in tuna meat observed by using X-ray computed tomography. *Int. J. Refrig.* **2015**, *60*, 270–277. [CrossRef]

15.   Sato, M.; Kajiwara, K.; Sano, N. Non-destructive three-dimensional observation of structure of ice grains in frozen food by X-ray computed tomography using synchrotron radiation. *Jpn. J. Food Eng.* **2016**, *17*, 83–88.

16.   Fischer, L. *An Introduction to Gel Chromatography*; North-Holland Publishing Company: Amsterdam, The Netherlands, 1969.

17.   Ruike, M.; Takada, S.; Murase, N.; Watanabe, T. Changes in the bead structure of crosslinked polymer gels during drying and freezing. *Cryo-Letters* **1999**, *20*, 61–68.

18.   Murase, N.; Yamada, S.; Ijima, N. Ice crystallization in gels and foods manipulated by the polymer network. In *Water Properties in Food, Health, Pharmaceutical and Biological Systems: ISOPOW 10*; Reid, D.S., Sajjaanantakul, T., Lillford, P.J., Charoenrein, S., Eds.; Wiley-Blackwell: Ames, IA, USA, 2010; pp. 373–383, ISBN 0-8138-1273-9.

gels

MDPI

Article

Analysis of Heterogeneous Gelation Dynamics and Their Application to Blood Coagulation

Toshiaki Dobashi [1,*] and Takao Yamamoto [2]

1   Division of Molecular Science, Graduate School of Science and Technology, Gunma University, Kiryu, Gunma 376-8515, Japan
2   Division of Pure and Applied Science, Graduate School of Science and Technology, Gunma University, Kiryu, Gunma 376-8515, Japan; tyam@gunma-u.ac.jp
*   Correspondence: dobashi@gunma-u.ac.jp; Tel.: +81-277-30-1427

Received: 11 May 2018; Accepted: 5 July 2018; Published: 9 July 2018

Abstract: We present a scaling model based on a moving boundary picture to describe heterogeneous gelation dynamics. The dynamics of gelation induced by different gelation mechanisms is expressed by the scaled equation for the time taken for development of the gel layer with a few kinetic coefficients characterizing the system. The physical meaning obtained by the analysis for a simple boundary condition from the standpoint of the phase transition shows that the time development of the gelation layer depends on whether the dynamics of the order parameter expressing the gelation of the polymer solution is fast or slow compared with the diffusion of the gelators in the heterogeneous gelation. The analytical method is used to understand the coagulation of blood from various animals. An experiment using systems with plasma coagulation occurring at interfaces with calcium chloride solution and with packed erythrocytes is performed to provide the data for model fitting and it is clarified that a few key kinetic coefficients in plasma coagulation can be estimated from the analysis of gelation dynamics.

Keywords: heterogeneous gelation dynamics; moving boundary picture; phase transition dynamics; kinetic coefficient; blood coagulation

1. Introduction

Commonly used gels are prepared using chemical reagents while mixing, by lowering the temperature and by the irradiation of high-energy electromagnetic waves, such as an electron beam and UV, where the cross-linking sites of polymer networks are dispersed randomly in gels, resulting in a macroscopically isotropic and homogeneous structure. This is because the maximum entropy of easily deformed polymer segments between cross-linking sites is obtained when the number of conformational states is maximum. Therefore, to prepare anisotropic gels some asymmetry, such as an asymmetric external force on the gel, should be required. To date, most studies performed on gelation have been on homogeneous systems [1–3]. In industry, however, gels are frequently prepared at an interface between two phases. Following the classification of chemical reactions into homogeneous and heterogeneous reactions, we refer to the gelation occurring in a homogeneous phase as homogeneous gelation and the gelation involving reactions at interfaces between different phases as heterogeneous gelation. In 1954, Thiele reported the fabrication of anisotropic beads by dripping an alginate solution in a bath of calcium chloride aqueous solution [4]. Thiele and coworkers prepared anisotropic channel-like alginate gels induced by diffusion of various multivalent cations [5]. These are among the first examples of heterogeneous gelation, on which many studies have now been reported [6–17]. The most important gels yielded by heterogeneous gelation are gels in parts of the human body. They are anisotropic, with the anisotropy often having a gradient, such as blood vessel walls [18]. Although the mechanism of biosynthesis is rather complex, most biotissues are yielded from a surface or a point and the growth is directional [19]. Therefore, understanding heterogeneous gelation is important from both industrial and biological viewpoints.

Among the systems that undergo heterogeneous gelation, polymer solutions in contact with solutions containing gelators such as cross-linkers have many promising applications. As a typical example of an anisotropic gel whose structure has been studied is a cylindrical gel prepared by the dialysis of sodium alginate aqueous solution in calcium chloride aqueous solution [4–10,20]. Maki et al., performed a small-angle X-ray scattering study on the structure of the alginate gel and showed that the polymer orientation is perpendicular to the flow of the gelators [20]. They showed that the gel is highly anisotropic near the surface (dialysis tube) but becomes isotropic toward the center. It has been shown for many combinations of polymers and gelators that the orientation is always similar to that observed for the alginate [14,15,21]. By changing the boundary conditions, we can prepare anisotropic gels with various shapes such as spherical gels and fiber-like gels [8]. The birefringence distribution of the gels is consistent with the orientation of polymer molecules perpendicular to the flow of the gelators [8]. In these systems, polymer molecules cannot permeate into the gel and the gelators selectively diffuse into the already yielded gel, resulting in the cross-linking of polymers at the reaction front. Since the polymers should have a high affinity with the gelators, they are in contact with the interface over a large contact area, which results in molecular orientation. Study of the microrheological events occurring on polymer molecules in the boundary layer at the sol-gel interface is challenging, although such studies should be performed in the near future.

For applications, the present approach to yielding anisotropy has an advantage of gelation in a self-organized manner under mild conditions. Several elaborate gels having unique properties have been developed. Gong's group at Hokkaido University used a double network gelation involving the contact of synthetic polymer solutions with gelator solutions to prepare a variety of tough anisotropic artificial substitutes such as artificial cartilage [22,23]. Furusawa et al. prepared epithelial lumen-like engineered tissues by the contact of collagen solutions with buffer solutions having a high pH due to spinodal decomposition [24]. Konno et al., prepared a multilayer cylindrical gel consisting of an inner isotropic gel layer and an outer highly anisotropic gel layer [25,26]. The release of drugs from gel particles having such multilayers may exhibit different behavior from the common exponential behavior observed for homogeneous particles and used for stepwise or constant drug release, similarly to capsules or spheres having layer-by-layer wall membranes [27].

On the other hand, the results of studies on the dynamics of heterogeneous gelation have not been organized and almost no applications of such gelation dynamics have been reported, although basic data on the gelation dynamics for simple polymer systems that undergo gelation in contact with gelator solutions have been accumulated [28–33]. Measurable physical quantities characterizing the gelation process are the gel fraction and the molecular weight distribution in the sol fraction in homogeneous gelation [1], whereas they are the gel volume and the degree of orientation of the gel in heterogeneous gelation [8]. In most cases, gel layers are formed at the interface between the polymer solution and the gelator solution and extended into the polymer solution while keeping the sol-gel interface clear [8]. Yamamoto et al., developed an irreversible thermodynamic theory focusing on the motion of the sol-gel boundary (moving boundary (MB) picture) [28,29].

In this article, we attempt to reorganize the experimental results on the basis of the MB picture, focusing on the role of the gelator. Then we propose a way of classifying the previously reported systems into several types as a first step toward understanding the characteristics of heterogeneous gelation dynamics. The gelation dynamics are expressed using universal scaling equations corresponding to the gelation mechanism, such as gelation induced by the inflow of cross-linkers (Case 1), by solvent exchange (Case 2), by the inflow of catalysts (Case 3), by the exchange of solutes having very different diffusion constants (Case 4) and by nucleation at low supersaturation (Case 5) (see Section 2.3 for details). In Section 3, we discuss the physical meaning of isotropic and anisotropic gelation from the standpoint of phase separation. By fitting data for gelation dynamics whose gelation mechanism is unknown to the universal scaling equation for Cases 1–5, we can identify the type of the gelation of the system. Furthermore, we can determine kinetic coefficients from the fitting parameters. The second aim of this article is to apply the analytical method to one of the most important biomedical processes, blood coagulation. Although some properties of the blood of patients can be obtained from conventional biochemical tests and the time required for blood coagulation, we have few means of estimating kinetic properties in blood coagulation, which is part of the missing link between blood properties and disorder in blood coagulation. Analysis of the dynamics of blood coagulation by fitting the data to the universal scaling equation enables us to extract information on kinetic coefficients in the process that cannot be obtained by static measurements. In Section 4, we describe the application of the analytical method for gelation dynamics to two model systems of blood coagulation to determine several key kinetic coefficients relating to blood coagulation.

2. Experimental Results on Gelation Dynamics and Classification of Systems by Moving Boundary Picture

We propose a generalized model of heterogeneous gelation and show the observed gelation dynamics of chitosan solution [30] as a typical example in Section 2.1. Then we demonstrate a theoretical analysis of the dynamics based on the MB picture in Section 2.2. In Section 2.3 we discuss the factors which determine the characteristics of the dynamics in Section 2.2 and propose a way of classifying of various systems according to the key factors.

2.1. Gelation Dynamics of Chitosan Solution in Contact with Solutions with High pH

The gel growth behaviors from the interface of polymer solutions with various types of gelator solutions have been observed. In Figure 1 we show an illustration of one of the simplest cases, the one-dimensional growth of a gel in a polymer solution cell in contact with a gelator solution bath (left-hand side of the cell). This model was proposed for the analysis of gelation dynamics of chitosan solution induced by a change of pH. Figure 2 shows a typical observed time course of the gel layer thickness $X(t)$ induced at the interface of chitosan solution with NaOH solution [25]. Chitosan solution is soluble at a low pH and forms a gel at a neutral pH by hydrogen bonding. Under the geometry shown in Figure 1, the part of the chitosan solution where the pH changes from a low pH to a neutral pH is transformed to a gel. The initial process appears to be expressed by square-root behavior $X \sim \sqrt{t}$, as shown in the inset of Figure 2, suggesting diffusion-limited dynamics. Here we modify the model shown in Figure 1 and generalize it by including various cases in which A and B have different roles other than those of NaOH solution and chitosan solution, respectively. For example, the outflow of B may or may not be involved in gelation and gelation can occur with or without the consumption of A in the generalized model. Similar initial behaviors have, however, been observed for different types of gelator solutions under various geometries, whereas the late-stage behaviors were different from each other [28–30].

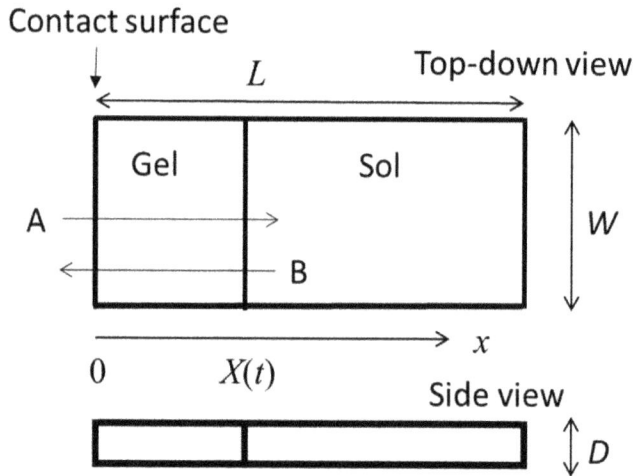

Figure 1. Illustration of one-dimensional gel layer growth from the contact surface between polymer solution enclosed in an L (length) × W (width) × D (depth) rectangular cell in an immersing gelator solution (left-hand side of the cell) induced by inflow of A and outflow of B. The x-axis is chosen to be perpendicular to the contact surface and is oriented in the direction from the immersing gelator solution to the polymer solution. The origin of the x-axis is chosen at the contact interface. $X(t)$ denotes the gel layer thickness at immersion time t. In the gelation of chitosan solution, the gelator solution and polymer solution are NaOH aqueous solution and chitosan in acetic acid aqueous solution, respectively.

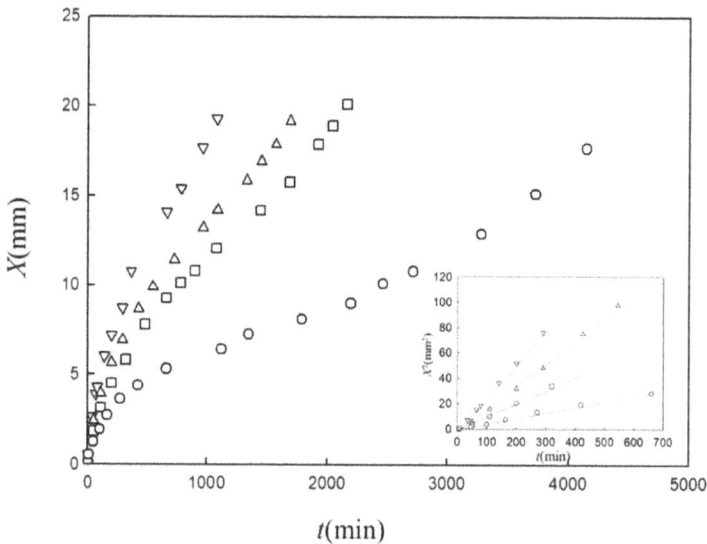

Figure 2. Time course of gel thickness X observed for one-dimensional growth of chitosan gel induced in 2 wt. % chitosan in 2 wt. % acetic acid aqueous solution in contact with sodium hydroxide aqueous solution with concentrations of 0.1 M (circles), 0.3 M (squares), 0.5 M (upward triangles) and 1 M (downward triangles) [25]. The inset shows the X^2 vs. t plot.

2.2. Moving Boundary Picture for Gelation Dynamics

The MB picture proposed by Yamamoto et al. [28,29] describes the gelation dynamics in terms of the sol-gel interface motion caused by the inflow of the gelator. The idea of the MB picture will be briefly explained through the gelation dynamics of chitosan solution. The dynamics introduced in Section 2.1 is analyzed as follows. In the system, the chitosan solution is enclosed in an L (length) × W (width) × D (depth) rectangular-solid space as shown in Figure 1 and is immersed in NaOH solution. The chitosan solution is in contact with the NaOH solution at a $W \times D$ side surface (the contact surface) [30].

The motion of the boundary $x = X(t)$ is derived from the following assumptions.

(A) The sodium ions flow into the sol part and the acetate ions flow out from the sol part to the NaOH solution through the gel layer. The neutralization caused by the flows instantly results in the cross-linking of the inner chitosan solution to produce a new gel layer.

(B) The gel layer does not capture the inflow sodium-ions and the outflow acetate-ions by acting as a sink and the flows change so slowly that they can be considered to be in a steady state; all the gelators inflowing from the gelator solution arrive at the inner polymer solution through the gel layer to realize a steady state.

(C) The NaOH solution, gel and inner sol are in local thermodynamic equilibrium at the boundaries.

The concentrations of sodium ions in the NaOH solution, in the inner sol, in the gel layer, at the interface between the gel and the NaOH solution at the position $x = x$ in the gel layer and at the interface between the gel and the inner sol in the gel layer are respectively denoted by ρ_s, ρ_0, $\rho(x)$, ρ_s' and ρ_0'. The concentrations of acetate ions at the corresponding positions are respectively denoted by C_s, C_0, $C(x)$, C_s' and C_0'. The chemical potentials of the sodium ion in the NaOH solution, in the inner sol and in the gel layer are respectively denoted by $\mu_{NaOH}^s(\rho_s)$, $\mu_{NaOH}^0(\rho_0)$ and $\mu_{NaOH}(\rho)$ and those of the acetate ion are denoted by $\mu_{Ac}^s(C_s)$, $\mu_{Ac}^0(C_0)$ and $\mu_{Ac}(C)$.

The flux of the inflow sodium-ion in the gel layer is expressed as

$$\vec{j}_{NaOH}(x) = j_{NaOH}(x)\vec{e}_x \tag{1}$$

where j_{NaOH} is the sodium ion flux density and is expressed in terms of the concentration $\rho(x)$ and velocity $v_{NaOH}(x)$ of the sodium ion as $j_{NaOH}(x) = \rho(x)v_{NaOH}(x)$. The unit vector along the x-axis is denoted by $\vec{e}_x$. According to Fick's law, the velocity of the sodium ion at x is given by

$$v_{NaOH}(x) = -k_{NaOH}\frac{\partial \mu_{NaOH}(\rho(x))}{\partial x} \tag{2}$$

where k_{NaOH} is the mobility of the sodium ion. Therefore, we have

$$j_{NaOH}(x) = -k_{NaOH}\rho(x)\frac{\partial \mu_{NaOH}(\rho(x))}{\partial x} \tag{3}$$

In a similar way, we have the flux of the outflow acetate-ion as

$$\vec{j}_{Ac}(x) = -j_{Ac}(x)\vec{e}_x \tag{4}$$

with

$$j_{Ac}(x) = k_{Ac}C(x)\frac{\partial \mu_{Ac}(C(x))}{\partial x} \tag{5}$$

In the above, k_{Ac} is the mobility of the acetate-ion.

Let a new gel layer thickness dX be produced during period dt by neutralization caused by the sodium ion inflow. Assumption (A) gives the following relationship:

$$\frac{1}{\rho_G(C_0(t))}j_{NaOH}(X)WDdt = WDdX \tag{6}$$

where $\rho_G(C_0)$ is the number of sodium ions required to neutralize a unit volume of the inner chitosan solution with acetate ion concentration C_0 and is reasonably assumed to be proportional to C_0,

$$\rho_G(C_0) = \alpha C_0 \tag{7}$$

where α is a positive constant. Note that C_0 is a function of the immersion time t. Hence, we have the time development equation for the gel thickness:

$$\frac{dX}{dt} = \frac{1}{\alpha C_0(t)}j_{NaOH}(X(t)) \tag{8}$$

Assumption (B) requires the following relationships:

$$\text{div}\,\vec{j}_{NaOH} = \frac{\partial j_{NaOH}}{\partial x} = -k_{NaOH}\frac{\partial}{\partial x}\left[\rho(x)\frac{\partial \mu_{NaOH}(\rho(x))}{\partial x}\right] = 0 \tag{9}$$

$$\text{div}\,\vec{j}_{Ac} = -\frac{\partial j_{Ac}}{\partial x} = -k_{Ac}\frac{\partial}{\partial x}\left[C(x)\frac{\partial \mu_{Ac}(C(x))}{\partial x}\right] = 0 \tag{10}$$

Integrating the differential Equations (9) and (10), we have the steady-state flows of sodium ions and acetate ions:

$$j_{NaOH} = k_{NaOH}\frac{g_{NaOH}(\rho_s') - g_{NaOH}(\rho_0')}{X} \tag{11}$$

$$j_{Ac} = -k_{Ac}\frac{g_{Ac}(C_s') - g_{Ac}(C_0')}{X} \tag{12}$$

where g_{NaOH} and g_{Ac} are the pressures of sodium ions and acetate ions respectively given by

$$g_{NaOH}(\rho) = \mu_{NaOH}(\rho)\rho - f_{NaOH}(\rho) \tag{13}$$

$$g_{Ac}(C) = \mu_{Ac}(C)C - f_{Ac}(C) \tag{14}$$

Here, the free energies f_{NaOH} and f_{Ac} per unit volume satisfy the relationships $\mu_{NaOH}(\rho) = \partial f_{NaOH}(\rho)/\partial\rho$ and $\mu_{Ac}(\rho) = \partial f_{Ac}(\rho)/\partial\rho$.

Assumption (C) gives the chemical potential balance

$$\begin{cases} \mu_{NaOH}(\rho_s') = \mu_{NaOH}^s(\rho_s) \\ \mu_{NaOH}(\rho_0') = \mu_{NaOH}^0(\rho_0) \\ \mu_{Ac}(C_s') = \mu_{Ac}^s(C_s) \\ \mu_{Ac}(C_0') = \mu_{Ac}^0(C_0) \end{cases} \tag{15}$$

Since the concentration of acetic acid is not so large, the chemical potential balance can be rewritten by assuming continuity of the concentration, $C_s' \cong C_s$ and $C_0' \cong C_0$. In the immersion solution, the concentration of acetic acid is very small, $C_s \cong 0$, because the volume of the immersion solution is very large. Hence, $C_s' \cong C_s \cong 0$. Therefore, the flux of the acetate-ion flow is given by

$$j_{Ac} = -k_{Ac}\frac{g_{Ac}(0) - g_{Ac}(C_0)}{X} \tag{16}$$

In the dilute limit of acetate ions in the gel, the pressure is expressed by $g_{Ac}(C) = k_B T C$ and the flux is given by

$$j_{Ac} = \frac{\beta C_0}{X} \tag{17}$$

where $\beta = k_{Ac} k_B T$. The acetate ion concentration in the sol part C_0 decreases with increasing immersion time t since acetate ions flow out from the sol part and the time development of the acetate ion concentration is given by

$$V(X)\frac{dC_0(t)}{dt} = -WDj_{Ac} \tag{18}$$

where $V(X) = WD(L - X)$ is the volume of the sol part. Using Equations (15) and (16), we have

$$\frac{dC_0(t)}{dt} = -\frac{\beta C_0(t)}{L - X(t)} \tag{19}$$

Treating the sodium ion flow in the same manner as the acetate ion flow, we have

$$j_{NaOH} = \frac{\gamma \rho_s}{X} \tag{20}$$

where $\gamma = k_{NaOH} k_B T$.

Solving the simultaneous equations, Equations (8), (19) and (20), with the initial condition $C_0(0) = C_{in}$ and $X(0) = 0$ and introducing the scaled variables $\tilde{X} = X/L$ and $\tilde{t} = t/L^2$, we have the scaled equation

$$\tilde{\zeta}\left(\tilde{X}, \frac{\beta}{K_{in}}\right) = K_{in}\tilde{t} \tag{21}$$

with

$$\tilde{\zeta}\left(\tilde{X}, \frac{\beta}{K_{in}}\right) = \int_0^{\tilde{X}} \frac{\tilde{u}}{1 - \frac{\beta}{K_{in}}\ln(1 - \tilde{u})} d\tilde{u} \tag{22}$$

and

$$K_{in} = \frac{\gamma}{\alpha}\frac{\rho_s}{C_{in}} \tag{23}$$

From the expansion $\tilde{\zeta}\left(\tilde{X}, \frac{\beta}{K_{in}}\right) = \frac{1}{2}\tilde{X}^2 + O\left(\tilde{X}^3\right)$ around $\tilde{X} = 0$, we derive the initial-stage ($\tilde{X} \approx 0$) behavior as

$$\frac{1}{2}\tilde{X}^2(\tilde{t}) = K_{in}\tilde{t} \tag{24}$$

This equation indicates square-root behavior in which the gel thickness increases proportionally to the square of the immersion time,

$$X = \sqrt{2K_{in}t} \tag{25}$$

The square-root behavior is a characteristic feature of the diffusion-limited dynamics.

The experimental results shown in Figure 2 are analyzed using Equations (21) and (22). In the analysis, K_{in} and β are the fitting parameters. The results are plotted according to the equations in Figure 3. The slope K_{in} is proportional to the NaOH concentration in the immersion solution, inversely proportional to the acetic acid concentration in the chitosan solution and independent of the chitosan concentration, as predicted by Equation (23) [30] (not shown). Therefore, the time course of the gel thickness is fully explained by the MB picture. The observed gelation behavior is expressed by the scaled Equation (21) with system-dependent coefficients K_{in} and β. Thus, we can obtain information on the kinetic coefficients γ and k_{Ac} of the system from the fitting parameters K_{in} and β by comparing the experimental data with the theoretical equation.

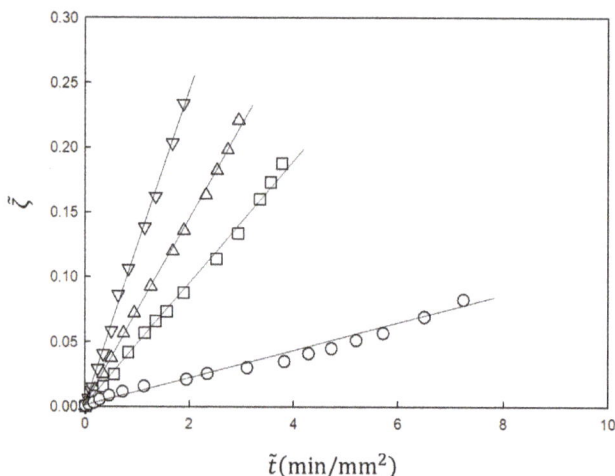

Figure 3. Time course of gel thickness obtained using the function ζ given by Equation (22) [30] for the data in Figure 2. ζ is proportional to $\tilde{t}$, as predicted by Equation (21).

2.3. Classification of Gelation Dynamics

In the gelation of chitosan solution, there are three types of characters having different roles. The first character is chitosan molecules, which are the element polymers constituting the gel. The second character is sodium ions, which are the gelator and are consumed to produce a gel layer. The third character is acetate ions, which are the gelation inhibitor.

Focusing on the gelator and the gelation inhibitor, we classify the gelation system into several types depending on the gelation mechanism. The simplest case is Case 1, in which cross-linking occurs simply by the inflow of cross-linkers as gelators; the gelation inhibitors are absent and the gelators are consumed to produce a gel layer. In Figure 1, the gelator A is involved in gelation whereas B is not involved in gelation. An example of Case 1 is a system where sodium alginate aqueous solution is in contact with calcium chloride aqueous solution (alginate/Ca^{2+} system) [4–10,20], in which calcium ions are the gelators. In Case 2, gelators flow into the element polymer solution and the gelation inhibitors flow out. Examples of Case 2 are polymers such as chitosan and collagen that undergo gelation via the formation of hydrogen bonds induced by a change in pH resulting from contact with a high pH solution. When chitosan or collagen in acetic acid solution with a low pH comes in contact with an aqueous solution of NaOH with a high or medium pH, anisotropic gels are prepared [15,30]. The gelator may be a catalyst, as in Case 3, making it necessary to consider the repeated use of the catalyst. In Case 3, although the inhibitors are absent, as in Case 1, the gelators are *not* consumed in the gelation. An example of Case 3 is the system where gelatin aqueous solution is in contact with transglutaminase aqueous solution [34].

Note that transient viscoelastic change occurs without any reactions when both solutes in liquid phases have very different diffusion constants [35]. We define this case as Case 4. In Case 4, the inhibitors are absent, as in Case 1. However, the gelator makes *no* links between the polymers. An example of Case 4 is the system where a high-molecular-weight DNA aqueous solution sandwiched between a pair of cover glasses is immersed in a low-molecular-weight DNA aqueous solution. In this case, the initial inflow of low-molecular-weight DNA causes the high-molecular-weight DNA solution to transiently form an anisotropic gel-like substance owning to the excluded volume effect and then the high-molecular-weight DNA diffuses to the immersion solution, finally resulting in a homogeneous solution. The gelation dynamics of the above-mentioned typical diffusion-limited systems were derived in Section 2.2 and are summarized in Table 1.

Table 1. Types of gelation and time development equation.

Case #	Geometry*	Time Development of Gel Thickness	System-Dependent Parameters	Equation #	Ref.
1	One-dimensional	$x = \sqrt{2Kt}$	$K = kk_BT\dfrac{\rho_S}{\rho_G}$	T1-1	
1	Cylindrically symmetrical	$\tilde{y} = \frac{1}{2}(1-\tilde{x})^2\ln(1-\tilde{x}) - \frac{1}{4}\tilde{x}^2 + \frac{1}{2}\tilde{x} = \tilde{K}\tilde{t}$	$K = kk_BT\dfrac{\rho_S}{\rho_G}$	T1-2	[26]
2	One-dimensional	$\tilde{\zeta}\left(\tilde{x},\frac{\beta}{K_{in}}\right) = \int_0^{\tilde{x}} \dfrac{\tilde{u}}{1 - \frac{\beta}{K_{in}}\ln(1-\tilde{u})}\,d\tilde{u} = K_{in}\tilde{t}$	$K_{in} = \dfrac{k_{NaOH}k_BT}{\alpha}\dfrac{\rho_S}{C_{in}}, \; \beta = k_{Ac}k_BT$	T2-1	[25]
2	Cylindrically symmetrical	$\tilde{z}\left(\tilde{x};\frac{2\beta}{K_{in}}\right) = \int_0^{\tilde{x}} \dfrac{(1-\tilde{u})\ln\frac{1}{1-\tilde{u}}}{1 - \frac{2\beta}{K_{in}}\ln(1-\tilde{u})}\,d\tilde{u} = K_{in}\tilde{t}$	$K_{in} = \dfrac{k_{NaOH}k_BT}{\alpha}\dfrac{\rho_S}{C_{in}}, \; \beta = k_{Ac}k_BT$	T2-2	[28]
3	One-dimensional	$x = \sqrt{2Kt}$	$K = 2k_{ct}k_BT$	T-3	
4	One-dimensional	$x = \sqrt{2Kt}$ (initial process)	$K = kk_BT\dfrac{\rho_S}{\rho_G}$	T1-1	

* One dimensional geometry corresponds to Figure 1. Cylindrically symmetrical geometry refers the case such as dialysis of polymer solutions to gelator solutions. x: gel thickness; $\tilde{x}$: gel thickness scaled by the radius of cylinder or the effective length of the cell; t: time; $\tilde{t}$: time scaled by square of the radius of cylinder or of the effective length of the cell; ρ_G: the number of gelator required to produce a unit volume of gel; ρ_S: concentration of gelator or acetate ion; C_{in}: initial concentration of sodium ion; α: a positive numerical factor; k: mobility of gelator; k_{NaOH}: mobility of sodium ion; k_{Ac}: mobility of acetate ion; k_{ct}: mobility of catalyst; k_B: Boltzmann constant; T: absolute temperature.

Case 5 is similar to the crystal growth from solutions at low supersaturation. Assume that the free energy of a polymer solution has a double minimum with different values. The lower and higher values correspond to gel phase and sol phase, respectively. If a gel region is nucleated by contact with a solid phase, then the polymer solution gels and the front line of the gel moves forward. In Case 5, neither gelators nor gelation inhibitors are present. The gelation is not diffusion-limited in this case. An example of Case 5 is the system where plasma is in contact with packed erythrocytes, which is discussed in detail in Section 4.2. The cases given above are not exhaustive but are those for which examples are reported [28–35].

3. Theoretical View of Heterogeneous Gelation from the Standpoint of Phase Transition Dynamics

Here we regard gelation as a phase transition and discuss the gelation dynamics in the context of the phase transition dynamics [36–38] for the polymer solution in contact with the gelator solution. Let us discuss the gelation system corresponding to Case 1 as the simplest case. The state of the polymer solution depends on the concentration ρ of gelators in the polymer solution. The state of the polymer solution is described in terms of the *order parameter* ψ, which expresses the degree of gelation; $\psi > 0$ when the polymer solution is in the gel state and $\psi = 0$ when it is in the sol state. Then, the *local free energy* of the polymer solution is a function of ρ and ψ,

$$f = f(\rho, \psi) \tag{26}$$

Let us consider a one-dimensional gel growth system, as shown in Figure 1. The free-energy functional expressing the whole of the polymer solution is given by

$$F(\{\rho\}, \{\psi\}) = \int \left[\frac{1}{2}\kappa_\rho \left(\frac{\partial \rho}{\partial x}\right)^2 + \frac{1}{2}\kappa_\psi \left(\frac{\partial \psi}{\partial x}\right)^2 + f(\rho(x), \psi(x)) \right] dx \tag{27}$$

where the positive constants κ_ρ and κ_ψ express the increase in free energy with respect to the space inhomogeneity of the gelator concentration and the degree of gelation, respectively. Since the concentration of gelators is a conserved quantity, the time development equation for the gelator is given by

$$\frac{\partial \rho(x, t)}{\partial t} = \text{div}\left[\Gamma_\rho \nabla \frac{\delta F}{\delta \rho(x, t)}\right] \tag{28}$$

where t denotes the elapsed time and Γ_ρ is the kinetic coefficient for ρ. Since ψ is an order parameter and is not a conserved quantity, the time development equation for ψ is given by

$$\frac{\partial \psi(x, t)}{\partial t} = -\Gamma_\psi \frac{\delta F}{\delta \psi(x, t)} \tag{29}$$

where Γ_ψ is the kinetic coefficient for ψ. The boundary conditions

$$\mu_G(\rho(0, t), \psi(0, t)) = \mu_S \tag{30}$$

is imposed, where $\mu_G(\rho(0, t), \psi(0, t))$ and μ_S are, respectively, the chemical potential of the gelator in the polymer solution at the liquid-liquid contact interface and that in the gelator solution. In terms of F, μ_G is generally given by

$$\mu_G(\rho(x, t), \psi(x, t)) = \frac{\delta F}{\delta \rho(x, t)} \tag{31}$$

One of the most characteristic features of the heterogeneous gelation process is that the boundary between the sol phase and gel phase is macroscopically clear. Thus, the gel growth dynamics can be visualized by the MB picture. The dynamics shown in Table 1 was derived by means of the MB picture.

Assumptions (A)–(C) in Section 2.2 for the gelation of chitosan solution (Case 2) is modified for Case 1 as follows [28] since the outflow ions are not involved in gelation in Case 1.

(A′) All the inflow gelators arriving at the inner polymer solution instantly cross-link the polymers to produce a new gel layer.

(B′) The gel layer does not capture the inflow gelators by acting as a sink and the gelator flow changes so slowly that it can be considered to be in a steady state; all the gelators inflowing from the gelator solution arrive at the inner polymer solution through the gel layer to realize a steady state.

(C′) The gelator solution, gel and inner polymer solution are in local thermodynamic equilibrium at the boundary.

According to the above three assumptions, in terms of the concentration $\rho(x)$ and chemical potential $\mu_G(\rho(x))$ of the gelator in the gel layer, the gelator flow density $j(x)$ is given by

$$j(x) = -k\rho(x)\frac{\partial \mu_G(\rho(x))}{\partial x} \tag{32}$$

where k is the mobility of the gelator and we modify Equations (8), (19) and (20) governing the time development equations for the sol-gel boundary $X(t)$ as [22]

$$\begin{cases} \frac{dX(t)}{dt} = \frac{j(X(t))}{\rho_G} \\ j(X) = kk_BT\frac{\rho_s}{X} \end{cases} \tag{33}$$

where ρ_G is the number of gelators required to produce a unit volume of gel. Note that ρ_G is constant and the equation corresponding to Equation (18) is absent in Case 1 since gelation inhibitors are absent.

Using the initial condition $X(0) = 0$, we obtain the solution of Equation (33) as

$$X(t) = \sqrt{2Kt} \tag{34}$$

where $K = kk_BT$. For Case 1, the square root behavior occurs during the entire period, although the behavior only occurs during the initial stage for Case 2.

Let us compare the above results with the phase transition dynamics given by Equations (28) and (29). The MB picture does not refer to the time development of the degree of gelation. Assumption (A′), however, requires that the time development of ψ is very rapid and an equilibrium value that depends on the gelator density ρ is quickly achieved. Thus, the degree of gelation ψ is expressed as

$$\psi(x,t) \approx \psi_0\Theta(X(t) - x) \tag{35}$$

where for simplicity we let the "level function" $\Theta(x)$ be given by the step function defined by

$$\Theta(x) = \begin{cases} 0 & x < 0 \\ 1 & x > 0 \end{cases} \tag{36}$$

and ψ_0 is a positive constant satisfying the equation

$$\frac{\partial f(\rho_G, \psi_0)}{\partial \psi_0} = 0 \tag{37}$$

The gelator concentration in the MB picture indicates the concentration $\delta\rho(x)$ of the gelator dissolved in the solvent, which is given by

$$\delta\rho(x) = \rho(x) - \rho_G \tag{38}$$

Therefore, in the context of phase transition dynamics, Equation (32) should be rewritten as

$$j(x) = -k\delta\rho(x)\frac{\partial\mu_G(\rho(x))}{\partial x} \tag{39}$$

Equation (39) requires the relationship

$$\Gamma_\rho = \Gamma_\rho(\rho,\psi_0) = k\delta\rho \tag{40}$$

for the kinetic coefficient $\Gamma_\rho = \Gamma_\rho(\rho,\psi)$. Assumption (B') requires that the gelator flow can be regarded as a steady flow. The time dependence of the gelator concentration ρ is very weak and appears only through the time dependence of the boundary $X(t)$; the boundary condition, that is, Assumption (C'), determines the weak time dependence. Hence, the gelator concentration is given by

$$\rho(x,t) \approx (\rho_G + \delta\rho(x;X(t)))\Theta(X(t) - x) \tag{41}$$

The chemical potential $\mu_G(x)$ in Equations (32) and (39) is given by

$$\mu_G(x) = \frac{\partial f(\rho_G + \delta\rho(x),\psi_0)}{\partial\delta\rho(x)} \tag{42}$$

To effectively use the description of the phase transition dynamics given by Equations (28) and (29), it is necessary to clarify the function forms of $f(\rho,\psi)$ and Γ_ρ and the value of the kinetic constant Γ_ψ statistical-mechanically and/or experimentally. This requirement will be discussed as a future issue. Here, we discuss a problem independent of the details of $f(\rho,\psi)$ and Γ_ρ. Equation (33), expressing the gelation dynamics, is valid when the dynamics of $\psi(x,t)$ is very fast compared with the gelator diffusion. What kind of dynamics is obtained in the opposite case, in which the dynamics of $\psi(x,t)$ is very slow? The answer is obtained from the common phase transition dynamics. In this case, Equations (28) and (29) are rewritten as

$$\frac{\partial f(\rho_{eq},0)}{\partial\rho_{eq}} = 0 \tag{43}$$

$$\frac{\partial\psi(x,t)}{\partial t} = -\Gamma_\psi\frac{\delta F_\psi}{\delta\psi(x,t)} \tag{44}$$

with

$$F_\psi = \int\left[\frac{1}{2}\kappa_\psi\left(\frac{\partial\psi}{\partial x}\right)^2 + f(\rho_{eq},\psi)\right]dx \tag{45}$$

The boundary condition is given by

$$\begin{cases} \psi(0,t) = \psi_{eq} \\ \lim_{x\to\infty}\psi(x,t) = 0 \end{cases} \tag{46}$$

where the equilibrium value of the degree of gelation ψ_{eq} is obtained from the equation

$$\frac{\partial f(\rho_{eq},\psi_{eq})}{\partial\psi_{eq}} = 0 \tag{47}$$

The dynamics described by Equation (44) is limited by the free energy and is regarded as the relaxation from an unstable state to a stable state minimizing the free energy.

When $f(\rho_{eq}, \psi)$ is a double-well-type function (both the sol phase ($\psi = 0$) and the gel phase ($\psi = \psi_{eq}$) locally minimize the local free energy $f(\rho_{eq}, \psi)$) and the constant κ_ψ is small, the solution of Equation (44) is approximately [37,39]

$$\psi(x, t) \approx \psi_{eq}\Theta(X(t) - x) \tag{48}$$

In this case, the gelation dynamics is also visualized by the motion of the sol-gel boundary. Therefore, to pay attention to the motion of the sol-gel boundary is valid. We could call the moving boundary picture the analysis method in which we pay attention to the sol-gel boundary motion even in the case of the energy-limited dynamics. The time dependence of $X(t)$, however, differs from the square-root behavior. The dynamics is expected to effectively visualize the gelation dynamics for Case 5.

Finally, we discuss how the anisotropy is taken into account. The presence of a gelator concentration gradient $\partial\rho(x)/\partial x$ during gelation is one of the characteristics of heterogeneous gelation. Therefore, we assume that the local free energy depends on the concentration gradient and introduce the order parameter ϕ expressing the degree of anisotropy of the polymer solution, where $\phi > 0$ when the polymer solution is anisotropic and $\phi = 0$ when it is isotropic. The local free energy is given by

$$f = f\left(\rho, \frac{\partial\rho}{\partial x}, \psi, \phi\right) \tag{49}$$

The free-energy functional is given by

$$F\left(\rho, \frac{\partial\rho}{\partial x}, \psi, \phi\right) = \int \left[\frac{1}{2}\kappa_\rho\left(\frac{\partial\rho}{\partial x}\right)^2 + \frac{1}{2}\kappa_\psi\left(\frac{\partial\psi}{\partial x}\right)^2 + \frac{1}{2}\kappa_\phi\left(\frac{\partial\phi}{\partial x}\right)^2 + f\left(\rho, \frac{\partial\rho}{\partial x}, \psi, \phi\right)\right] dx \tag{50}$$

where the positive constant κ_ϕ expresses the increase in free energy with respect to the space inhomogeneity of the anisotropy. In addition to Equations (28) and (29), we introduce the time development equation for ϕ as

$$\frac{\partial\phi(x, t)}{\partial t} = -\Gamma_\phi \frac{\delta F}{\delta\phi(x, t)} \tag{51}$$

where Γ_ϕ is the kinetic coefficient for ϕ.

Let us discuss the diffusion-limited gelation dynamics with the anisotropy expressed by the MB picture. The gelation dynamics of curdlan solution [28] is in this category. In this case, the time development of the anisotropy is expressed as

$$\phi(x, t) \approx \phi_0\Theta(X(t) - x) \tag{52}$$

where

$$\frac{\partial f\left(\rho_G, \frac{\partial\delta\rho(X)}{\partial X}, 0, \phi_0\right)}{\partial\phi_0} = 0. \tag{53}$$

Equation (53) requires the local free energy to reach a local minimum at a finite value of the degree of anisotropy. Note that global minimization of the local free energy at a finite ϕ is not required. At a concentration gradient $\left|\frac{\partial\delta\rho(X)}{\partial X}\right|$ larger than a threshold value, the local free energy is expected to have such a local minimum. The increase in the degree of gelation roughly coincides with the increase in anisotropy. The kinetic coefficient Γ_ϕ rapidly drops to zero with increasing ψ since the gelation significantly interferes with the polymer motion in the polymer solution. Thus, the degree of anisotropy is fixed at a finite value ϕ_0.

4. Application of Scaled Gelation Dynamics to Analysis of Blood Coagulation

One of the most interesting applications of the analysis of heterogeneous gelation dynamics is blood coagulation (gelation). Blood consists of about 45% blood corpuscles, which are mainly erythrocytes and the remaining liquid component, plasma. Fibrinogens, which comprise 7% of blood protein, are the main component of coagulants (gels). Blood coagulation is triggered by the contact of plasma with coagulant factors at cell surfaces or blood vessel walls, as illustrated in Figure 4. Therefore, the main process of macroscopic blood coagulation is the gelation of plasma induced by the contact of plasma and coagulant factors at a cell surface. With a sufficient amount of calcium ions, after complex cascade reactions, the key protein thrombins are activated and fibrinogens are hydrolyzed to form fibrins and then protofibrils, which are the building blocks of the coagulant. The biochemical reaction cascade in blood coagulation has been established by considering a number of simplified homogeneous systems [40]. However, the research on the dynamic aspects of blood coagulation by considering heterogeneous systems, such as a thrombin solution/plasma contact system, began relatively recently [41,42]. Determination of the kinetic coefficients of blood coagulation should help provide the missing link between the biochemical properties and disorders in blood coagulation. The kinetic coefficients can potentially be used as direct indicators of blood coagulability in diagnosis.

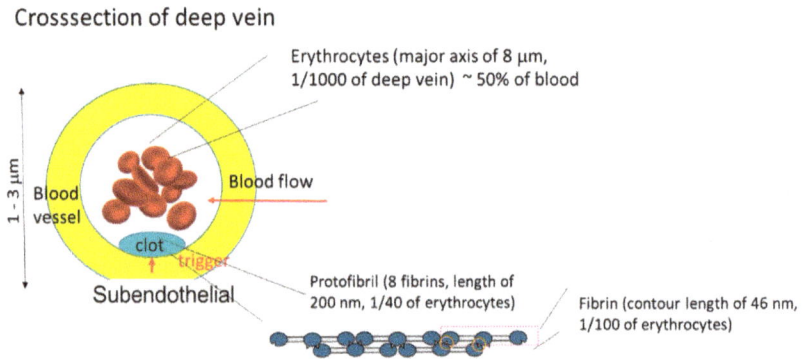

Figure 4. Illustration of blood coagulation at blood vessel wall.

It is very difficult to completely replace an in vivo system by an in vitro system for blood testing, because when we draw blood from a body we must add some anticoagulants such as citrates or heparin to the blood. Therefore, it is important to study key systems in which part of the blood coagulation process can be reproduced to obtain useful information on blood. Here we discuss two systems of interest, citrated plasma in contact with calcium chloride aqueous solution (plasma/Ca$^+$) and citrated plasma in contact with packed erythrocytes (plasma/packed erythrocytes).

4.1. Analysis of Citrated Plasma/Ca$^+$ Contact System

In the clinical test to assess blood coagulability, a trigger protein or trigger lipid and calcium ions are mixed with citrated plasma and the clotting time is measured and compared with the standard value. Let us consider heterogeneous gelation as a control clinical testing system; this system consists of citrated plasma in contact with a calcium ion solution separated by a dialysis membrane, as illustrated in Figure 1. There is an inflow of calcium ions and an outflow of citrate ions through the dialysis membrane. When the calcium ion concentration in the plasma exceeds the threshold, gelation occurs. If we replace citrate ions, calcium ions and fibrinogen by acetate ions, sodium ions and chitosan, respectively, this system corresponds to Case 2 [43]. Therefore, the measurement of the gel layer thickness X as a function of time after the liquid-liquid contact t can be compared with the theoretical equation, Equation (T2-1) in Table 1, of

$$\tilde{\zeta}\left(\tilde{X}, \frac{\beta}{K_{in}}\right) = \int_0^{\tilde{X}} \frac{\tilde{u}}{1 - \frac{\beta}{K_{in}}\ln(1-\tilde{u})}d\tilde{u} = K_{in}\tilde{t} \tag{54}$$

where

$$\tilde{X} = X/L, \quad \tilde{t} = t/L^2 \tag{55}$$

and L is the effective length of the cell. Since the blood coagulation dynamics is expressed by the scaled form of Equation (54), the analytical results based on the equation do not depend on the experimental tools. This is an advantage of this method of extracting valuable information on blood.

Recently the entire time course of the gel thickness was compared with the theory for bovine blood and was found to be well expressed by Equation (54) [43]. However, since the blood coagulation behavior of animals of the even-hoof class, such as cattle, is known to be considerably different from that of other animals such as human, swine and horses [44], in this paper we performed the corresponding experiment using blood from a horse, one of the animals of the odd-hoof class. The time course was well expressed by Equation (54) as shown in Figure 5. The scaled time lag k used for fitting was close to the clotting time, which was the time required for macroscopic gelation by mixing (homogeneous gelation), used in clinical testing. Therefore, the analysis is also applicable for the blood of animals other than even-hoof animals. Note that the observed data could only be fit to Case 2 among the five cases. Calcium ions are known to play various roles in blood. They bind plasma proteins that are both involved and not involved in coagulation, each with a different binding constant. For example, fibrinogen has several strong and weak binding sites with calcium ions. Some enzymes are only activated in the presence of a sufficient amount of calcium ions. Thus, blood coagulation is generally complex. The present system is regarded as a simplified one to relate the observed gelation dynamics to Case 2. Here, we further expand the expressions for parameters β and K_{in}. From the definition, we have

$$\beta = \frac{\alpha\beta_0 C_{in}}{\rho_G^0 + \alpha C_{in}} \tag{56}$$

$$\beta_0 = k_{CIT}k_B T \tag{57}$$

$$K_{in} = \frac{\gamma\rho_s}{\rho_G^0 + \alpha C_{in}} \tag{58}$$

$$\gamma = k_{Ca}k_B T \tag{59}$$

where C_{in} is the initial concentration of citrate in the plasma, ρ_s is the initial concentration of calcium ions in the calcium chloride aqueous solution, γ and β_0 are the diffusion constants of calcium ions and citrate ions in the gel, respectively and ρ_G^0 is the calcium ion concentration required for the gelation of a unit volume of plasma. From Equations (56)–(59), we have

$$C_{in} = -\rho_G^0 + \gamma\frac{\rho_s^0}{K_{in}} \tag{60a}$$

and

$$C_{in} = -\rho_G^0 + \beta_0\frac{C_{in}}{\beta} \tag{60b}$$

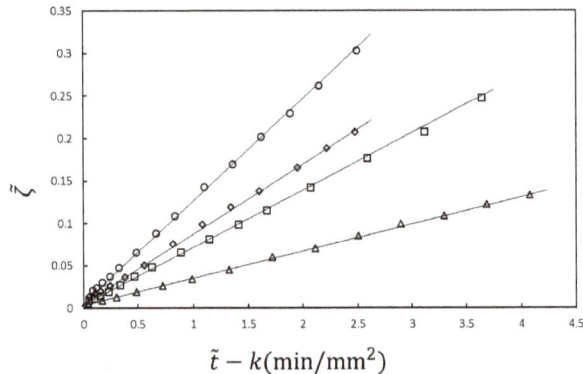

$$\tilde{t} - k(\mathrm{min/mm^2})$$

Figure 5. Gelation dynamics expressed by Equation (54) for horse plasma in contact with calcium chloride with concentrations of 10 mM ($\triangle$), 30 mM ($\square$), 50 mM ($\lozenge$) and 100 mM ($\bigcirc$).

Therefore, by measuring the citrate concentration dependence of the parameters β and K_{in}, we can determine the three physically defined quantities γ, β_0 and $\rho_G{}^0$ from the plot using Equations (60a) and (60b). According to preliminary experiments, this analysis is also valid for human blood and the kinetic coefficients depends on subjects. Therefore, these parameters could be used for diagnosis as new indicators reflecting blood coagulability.

4.2. Analysis of Plasma/Packed Erythrocyte Contact System

Until recently, erythrocytes were regarded as being unrelated to blood coagulation. However, in the past few decades, several pieces of indirect evidence that erythrocytes have an active function in thrombosis and hemostasis have been found [45]. In particular, in the case of blood flow stagnation, such as in deep vein thrombosis, also called economy-class syndrome, the erythrocyte surface is suspected to be one of the factors that trigger blood coagulation [46]. Deep vein thrombosis is a blood clot that develops within a deep vein in the body, usually in the leg. When part of the blood clot breaks off, enters the bloodstream and blocks one of the blood vessels in the lungs, it leads to complications such as pulmonary embolism [47]. It has been assumed that protein C or protein S, both of which inhibit the coagulation cascade in the normal state, is inactivated, resulting in blood coagulation [48]. Kaibara et al. proposed a new pathway involving the activation of an intrinsic pathway in the coagulation cascade by the protein on the erythrocyte membrane, erythroelastase [46]. Although few continuous studies have been carried out on this pathway, the system involving the contact of erythrocytes and plasma seems to be appropriate for studying whether or not this scheme is consistent with or directly related to deep vein thrombosis. Because of the deformability of erythrocytes, we can easily prepare packed erythrocytes with a flat surface by centrifugation. In this study, we performed a study on gelation dynamics induced in swine plasma in contact with packed erythrocytes. Figure 6 shows the time course of the thickness of a gel grown from a packed erythrocyte surface. After a time lag, the gel thickness increased proportionally to time. The slope was roughly constant, irrespective of the calcium concentration, which means that the gelation behavior is the same above a threshold calcium concentration after gelation is initiated. According to the mechanism proposed by Kaibara et al., a coagulation protein called factor IX is activated on the surface of erythrocytes [46]. Then the coagulation cascade should be successively activated and positive feedback may yield a locally activated space with saturated fibrin molecules similarly to in vivo [49]. The fibrin molecules form a network from the surface of the packed erythrocytes, similarly to in the case of crystal growth, as illustrated in Figure 7. This situation is similar to Case 5. Choosing a quartic function of ψ in the function $f(\rho_{eq}, \psi)$ and the boundary condition Equation (46), we obtain the solution of Equation (48) according to Chan [50]. The position of the sol-gel interface is expressed by the linear function of time

$$X(t) = v(t - t_0) \tag{61}$$

where t_0 is the lag time required so that the boundary condition is satisfied and v is a positive constant depending on the free-energy difference $f(\rho_{eq}, \psi_{eq}) - f(\rho_{eq}, 0)$. Therefore, the observed linear increase in X with time t shown in Figure 6 is explained by Equation (61) and the proportionality coefficient is related to the free-energy difference between the initial state and the equilibrium state. On the other hand, such a gel layer was hardly observed on hardened erythrocytes modified by glutaraldehyde whose plasma membrane was inactivated. Thus, the experimental results are consistent with the mechanism proposed by Kaibara et al. [46]. The rate of increase in the gel layer thickness provides information on the coagulability of the examined blood through the proportionality coefficient v. According to a preliminary experiment, this approach is valid for human blood. The kinetic coefficient determined by the observed heterogeneous gelation dynamics may be related to the coagulability of the blood of subjects and can potentially be used as an indicator of deep vein thrombosis.

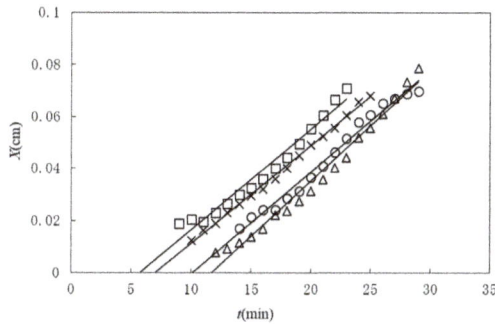

Figure 6. Time course of gel layer thickness observed on packed swine erythrocytes in contact with plasma with calcium chloride concentrations of 1.63 mM (○), 2.50 mM (△), 4.59 mM (□) and 6.68 mM (×).

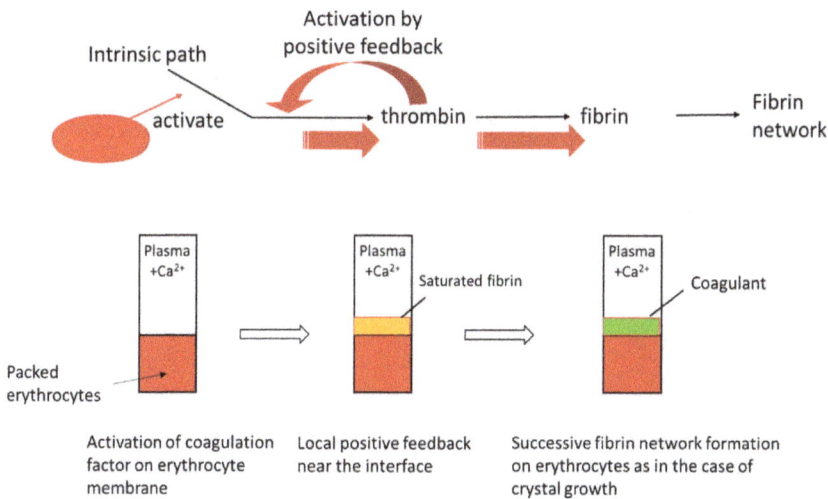

Figure 7. Scheme of coagulation induced by contact of plasma with packed erythrocytes.

5. Conclusions

The gelation dynamics induced by the contact of polymer and gelator solutions was classified into several types and expressed by a system-independent scaling equation and system-dependent coefficients. From the viewpoint of the phase transition, it was clarified that the time taken for the development of the gelation layer depends on whether the dynamics of the order parameter expressing the gelation of the polymer solution is fast or slow compared with the diffusion of the gelators. As a straightforward application of one type of gelation dynamics (Case 2), in vitro blood coagulation was analyzed to find kinetic coefficients characterizing the dynamics. New information on blood from human subjects can be extracted by the analysis of gel growth behavior. Heterogeneous gelation is involved in many industrial processes such as microencapsulation by an insolubilization reaction to prepare artificial eggs and drug delivery carriers. Information on the extent of the reaction is important for controlling the retention and release capability of the core substance. For spherical entities, however, it is difficult to trace the gelation dynamics. The theoretical analysis in Section 2 may also be applicable to these materials to enabling the gel layer thickness to be estimated at a desired time.

6. Materials and Method

To study the plasma/Ca$^+$ contact system, horse blood aseptically collected in equal volumes of Alsever's solution was purchased from Nippon Bio-test Laboratories Inc. The blood was centrifuged at 1600× g for 10 min at 25 °C to obtain platelet-free plasma (PFP). The plasma was poured in a cell (10 mm in height, 3 mm in width and 20 mm depth) made of poly(methylmethacrylate) from the bottom to the upper end (inlet). Then the inlet was sealed with a dialysis membrane (Sanko Jun-yaku Co., Ltd., Tokyo, Japan) to confine plasma proteins such as fibrinogen in the plasma. The cell was immersed in 40 mL of calcium chloride aqueous solution with various calcium chloride concentrations in the range of 10–100 mM. The motion of the boundary was observed and recorded with a digital camera at 25 °C.

To study the plasma/packed RBC contact system, swine blood provided by Gunma Meat Wholesale Market Co. Ltd. was citrated with 3.2% sodium citrate in a polypropylene tube. The suspension was centrifuged at 1600× g for 10 min at 25 °C to separate the PFP and packed erythrocytes. Hardened erythrocytes were prepared by mixing 20% erythrocyte suspension with 1.6% glutaraldehyde aqueous solution with 40 times the amount of erythrocyte suspension and gently stirring it for 72 h. Then the precipitates were rinsed with MilliQ water five times and physiological saline solution twice. After placing 2 mL of the PFP with the desired concentration of calcium ions on 1 mL of packed erythrocytes in a collagen-coated polystyrene tube, we observed the growth of a turbid layer from the interface over time at 25 °C. The volume of the turbid layer was estimated by image analysis and the gel volume was measured directly; these two volumes were found to be equivalent. Therefore, the growth of the gel layer was traced by observing the thickness of the turbid layer.

Author Contributions: T.D. designed the experimental study and T.Y. built the theory. T.D. and T.Y. wrote the paper.

Funding: This work was partly supported by JSPS KAKENHI Grant Numbers JP24350115, JP15K05241 and JP16H04031.

Acknowledgments: The authors are grateful to Yoshiharu Toyama for his helpful discussion and Ryuta Kurasawa and Hiroki Shinoda at Gunma University for their help with the blood coagulation experiments.

Conflicts of Interest: The authors declare no conflict of interest.

References

1. Flory, P.J. *Principles of Polymer Chemistry*; Cornell University Press: Ithaca, NY, USA, 1953.
2. Weiss, R.G.; Terech, P. *Molecular Gels Materials with Self-Assembled Fibrillar Networks*; Springer: Berlin, Germany, 2006.
3. Djabourov, M.; Nishinari, K.; Ross-Murphy, S.B. *Physical Gels from Biological and Synthetic Polymers*; Cambridge University Press: Cambridge, UK, 2013.

4. Thiele, H. Ordered coagulation and gel formation. *Discuss. Faraday Soc.* **1954**, *18*, 294–314. [CrossRef]
5. Thiele, H.; Hallich, K. Kapillarstrukturen in ionotropen Gelen. *Kolloid Z.* **1957**, *151*, 1–12.
6. Mørch, Y.A.; Donati, I.; Strand, B.L.; Skjåk-Bræk, G. Effect of Ca^{2+}, Ba^{2+} and Sr^{2+} on alginate microbeads. *Biomacromolecules* **2006**, *7*, 1471–1480. [CrossRef] [PubMed]
7. Thumbs, J.; Kohler, H.H. Capillaries in alginate gel as an example of dissipative structure formation. *Chem. Phys.* **1996**, *208*, 9–24. [CrossRef]
8. Dobashi, T.; Yamamoto, T. Anisotropic gel formation induced by dialysis. In *Encyclopedia of Biocolloid and Biointerface Science*; Ohshima, H., Ed.; Wiley: Hoboken, NJ, USA, 2016.
9. Schuster, E.; Sott, K.; Ström, A.; Altskär, A.; Smisdom, N.; Gebäck, L.N.; Hermansson, A.M. Interplay between flow and diffusion in capillary alginate hydrogels. *Soft Matter* **2016**, *12*, 3897–3907. [CrossRef] [PubMed]
10. Despang, F.; Dittrich, R.; Gelinsky, M. Novel biomaterials with parallel aligned pore channels by directed ionotropic gelation of alginate: Mimicking the anisotropic structure of bone tissue. In *Advances in Biomimetics*; George, A., Ed.; InTech: London, UK, 2011; Chapter 17; pp. 349–372.
11. Dobashi, T.; Nobe, M.; Yoshihara, H.; Yamamoto, T.; Konno, A. Liquid crystalline gel with refractive index gradient of curdlan. *Langmuir* **2004**, *20*, 6530–6534. [CrossRef] [PubMed]
12. Narita, T.; Tokita, M. Liesegang pattern formation in κ-carrageenan gel. *Langmuir* **2006**, *22*, 349–352. [CrossRef] [PubMed]
13. Rivas-Araiza, R.; Alcouffe, P.; Rochas, C.; Montembault, A.; David, L. Micron range morphology of physical chitosan hydrogels. *Langmuir* **2010**, *26*, 17495–17504. [CrossRef] [PubMed]
14. Mredha, M.T.I.; Zhang, X.; Nonoyama, T.; Nakajima, T.; Kurokawa, T.; Takagi, Y.; Gong, J.P. Swim bladder collagen forms hydrogel with macroscopic superstructure by diffusion induced fast gelation. *J. Mater. Chem. B* **2015**, *3*, 7658–7666. [CrossRef]
15. Furusawa, K.; Sato, S.; Masumoto, J.; Hanazaki, Y.; Maki, Y.; Dobashi, T.; Yamamoto, T.; Fukui, A.; Sasaki, N. Studies on the formation mechanism and the structure of the anisotropic collagen gel prepared by dialysis-induced anisotropic gelation. *Biomacromolecules* **2012**, *13*, 29–30. [CrossRef] [PubMed]
16. Mikkelsen, A.; Elgsaeter, A. Density ditribtuion of calcium-induced alginate gels. A numerical study. *Biopolymers* **1994**, *3*, 17–41.
17. Miyamoto, Y.; Kaysser, W.A.; Rabin, B.H.; Kawasaki, A.; Ford, R.G. (Eds.) *Functionally Graded Materials: Design, Processing and Applications*; Springer: Berlin, Germany, 1999.
18. Oka, S. *Cardiovascular Hemorheology*; Cambridge University Press: Cambridge, UK, 1981.
19. Alberts, B.; Johnson, A.; Lewis, J.; Raff, M.; Roberts, K.; Walter, P. *Molecular Biology of the Cell*, 5th ed.; Garland Science: New York, NY, USA, 2008.
20. Maki, Y.; Ito, K.; Hosoya, N.; Yoneyama, C.; Furusawa, K.; Yamamoto, T.; Dobashi, T.; Sugimoto, Y.; Wakabayashi, K. Anisotropic structure of calcium-induced alginate gels by optical and small-angle X-ray scattering measurements. *Biomacromolecules* **2011**, *12*, 2145–2152. [CrossRef] [PubMed]
21. Maki, Y.; Furusawa, K.; Yasuraoka, S.; Okamura, H.; Hosoya, N.; Sunaga, M.; Dobashi, T.; Sugimoto, Y.; Wakabayashi, K. Universality and specificity in molecular orientation in anisotropic diffusion method. *Carbohydr. Polym.* **2014**, *108*, 118–126. [CrossRef] [PubMed]
22. Yang, W.; Furukawa, H.; Gong, J.P. Highly extensible double-network gels with self-assembling anisotropic structure. *Adv. Mater.* **2008**, *20*, 4499–4503. [CrossRef]
23. Mredha, M.T.I.; Kitamura, N.; Nonoyama, T.; Wada, S.; Goto, K.; Zhang, X.; Nakajima, T.; Kurokawa, T.; Takagi, Y.; Yasuda, K.; et al. Anisotropic tough double network hydrogel from fish collagen and its spontaneous in vivo bonding to bone. *Biomaterials* **2017**, *132*, 85–95. [CrossRef] [PubMed]
24. Furusawa, K.; Mizutani, T.; Machino, H.; Yahata, S.; Fukui, A.; Sasaki, N. Application of multichannel collagen gels in construction of epithelial lumen-like engineered tissues. *ACS Biomater. Sci. Eng.* **2015**, *1*, 539–548. [CrossRef]
25. Konno, A.; Tsubouchi, M. Gel formation of curdlan. *Kinran Tanki Daigaku Kenkyushi* **1998**, *29*, 89–95.
26. Maki, Y.; Furusawa, K.; Dobashi, T.; Sugimoto, Y.; Wakabayashi, K. Small-angle X-ray and light scattering analysis of multi-layered curdlan gels prepared by a diffusion method. *Carbohydr. Polym.* **2017**, *155*, 136–145. [CrossRef] [PubMed]
27. Ariga, K.; Lvov, Y.M.; Kawakami, K.; Ji, Q.; Hill, J.P. Layer-by-layer self-assembled shells for drug delivery. *Adv. Drug Deliv. Rev.* **2011**, *63*, 762–771. [CrossRef] [PubMed]

28. Nobe, M.; Dobashi, T.; Yamamoto, T. Dynamics in dialysis process for liquid crystalline gel formation. *Langmuir* **2005**, *21*, 8155–8160. [CrossRef] [PubMed]
29. Yamamoto, T.; Tomita, N.; Maki, Y.; Dobashi, T. Dynamics in the process of formation of anisotropic chitosan hydrogel. *J. Phys. Chem. B* **2010**, *114*, 10002–10009. [CrossRef] [PubMed]
30. Rokugawa, I.; Tomita, N.; Dobashi, T.; Yamamoto, T. One-dimensional growth of hydrogel by a contact of chitosan solution with high-pH solution. *Soft Mater.* **2014**, *12*, 36–41. [CrossRef]
31. Skjåk-Bræk, G.; Grasdalen, H.; Smisrød, O. Inhomogeneous polysaccharide ionic gels. *Carbohydr. Polym.* **1989**, *10*, 31–54. [CrossRef]
32. Furusawa, K.; Minamisawa, Y.; Dobashi, T.; Yamamoto, T. Dynamics of liquid crystalline gelation of DNA. *J. Phys. Chem. B* **2007**, *111*, 14423–14430. [CrossRef] [PubMed]
33. Furusawa, K.; Narazaki, Y.; Tomita, N.; Dobashi, T.; Sasaki, N.; Yamamoto, T. Effect of pH on anisotropic gelation of DNA induced by aluminum cations. *J. Phys. Chem. B* **2010**, *114*, 13923–13932. [CrossRef] [PubMed]
34. Yamamoto, T.; Kakinoki, K.; Maki, Y.; Dobashi, T. Gelation and orientation dynamics of protein solution induced by enzyme solution. In Proceedings of the JPS 2016 Autumn Meeting, Kanazawa, Japan, 13 September 2016. 13aBE3.
35. Dobashi, T.; Maki, Y.; Furusawa, K.; Yamamoto, T. Anisotropic Structure Formation Induced by Liquid-Liquid Phase Contact and Diffusion. In Proceedings of the IUMRS-ICAM 2017, Kyoto, Japan, 31 August 2017. C6-I31-003.
36. Hohenberg, P.C.; Halperin, B.I. Theory of dynamic critical phenomena. *Rev. Mod. Phys.* **1977**, *49*, 437–479. [CrossRef]
37. Gunton, J.D.; Miguel, M.S.; Sahni, P.S. The Dynamics of First-order Phase Transitions. In *Phase Transitions and Critical Phenomena*, 8th ed.; Domb, C., Lebowitz, J.L., Eds.; Academic Press: London, UK; New York, NY, USA, 1983; pp. 267–466.
38. Hohenberg, P.C.; Krekhov, A.P. An introduction to the Ginzburg-Landau theory of phase transitions and nonequilibirm patterns. *Phys. Rep.* **2015**, *572*, 1–42. [CrossRef]
39. Allen, S.M.; Cahn, J.W. A microscopic theory for antiphase boundary motion and its application to antiphase domain coarsening. *Acta Metall.* **1979**, *27*, 1085–1095. [CrossRef]
40. Voet, D.; Voet, J.G. *Fundamentals of Biochemistry, Life at the Molecular Level*, 4th ed.; Wiley: Hoboken, NJ, USA, 2013.
41. Ataullakhanov, F.I.; Guria, G.T.; Sarbash, V.I.; Volkova, R.I. Spatiotemporal dynamics of clotting and pattern formation in human blood. *Biochim. Biophy. Acta* **1998**, *1425*, 453–468. [CrossRef]
42. Zhalyalov, A.S.; Panteleev, M.A.; Gracheva, M.A.; Ataullakhanov, F.I.; Shibeko, A.M. Co-ordinated spatial propagation of blood plasma clotting and fibrinolytic fronts. *PLoS ONE* **2017**, *12*, e0180668. [CrossRef] [PubMed]
43. Shida, N.; Kurasawa, R.; Maki, Y.; Toyama, Y.; Dobashi, T.; Yamamoto, T. Coagulation of plasma induced by a contact with calcium chloride solution. *Soft Matter* **2016**, *12*, 9471–9476. [CrossRef] [PubMed]
44. Kaibara, M.; Shinozaki, T.; Kita, R.; Iwata, H.; Ujiie, H.; Sasaki, K.; Li, J.Y.; Sawasaki, T.; Ogawa, H. Analysis of coagulation of blood in different animal species with special reference to procoagulant activity of red blood cell. *J. Jpn. Soc. Biorheol.* **2006**, *20*, 35–43.
45. Litvinov, I.; Weisel, J.W. Role of red blood cells in haemostasis and thrombosis. *ISBT Sci. Ser.* **2017**, *12*, 176–183. [CrossRef] [PubMed]
46. Iwata, I.; Kaibara, M.; Dohmae, N.; Takio, K.; Himeno, R.; Kawakami, S. Purification, identification and characterization of elastase on erythrocyte membrane as factor IX-activating enzyme. *Biochem. Biophys. Res. Commun.* **2004**, *316*, 65–70. [CrossRef] [PubMed]
47. Signorelli, S.S.; Ferrante, M.; Gaudio, A.; Fiore, V. Deep vein thrombosis related to environment. *Mol. Med. Rep.* **2017**, *15*, 3445–3448. [CrossRef] [PubMed]
48. Tapson, V.F. Acute pulmonary embolism. *N. Eng. J. Med.* **2008**, *358*, 1037–1052. [CrossRef] [PubMed]
49. Ieko, M. Characteristics of various anticoagulants and evaluation methods for risk of bleeding and thrombosis. *Jpn. J. Electrocardiol.* **2014**, *34*, 149–156.
50. Chan, S.-K. Steady-state kinetics of diffusion less first order phase transition. *J. Chem. Phys.* **1977**, *67*, 5755–5762. [CrossRef]

Article

Effect of Sweeteners on the Solvent Transport Behaviour of Mechanically-Constrained Agarose Gels

Isamu Kaneda

Food Physical Chemistry lab, Rakuno Gakuen University, Ebetsu, Hokkaido 069-8501, Japan; kaneda-i@rakuno.ac.jp; Tel.: +81-11-388-4701

Received: 27 February 2018; Accepted: 14 March 2018; Published: 16 March 2018

Abstract: Investigating the solvent transport behaviour of edible gels is important because it is strongly related to flavour release. We previously reported the solvent transport behaviour of mechanically-constrained agarose gels. These studies clearly showed that agarose gels can be treated as soft porous bodies. Herein, we investigated the effect of sweeteners on the solvent transport speed, which is an important issue in the food industry, using sucrose and xylitol. Sucrose caused a concentration-dependent reduction in solvent transport speed. One of the reasons for the effect is that the solvent to which sucrose was added reduced solvent flow speed within the porous agarose network. This finding provides valuable information for flavour release from compressed gels. Moreover, we found a similar effect for xylitol, which is a promising candidate for substituting sucrose in low-calorie foods. This study would provide basic knowledge for the development of a new type of low-calorie foods.

Keywords: agarose gel; compression; solvent transport; sucrose; xylitol

1. Introduction

Agarose is a neutral polysaccharide derived from Rhodophyta with a structure consisting of repeating 1,3-binding β-D-galactose and 1,4-binding 3,6-anhydro-α-L-galactose units. Agarose is insoluble in water at low temperatures owing to its double helix structure [1], but dissolves at high temperatures due to dissociation of the double helix. As agarose dissolved in water at high temperatures is allowed to cool and the thermal motion of the polymer chain is restrained, hydrogen bonds reform to create the double helix structure. If the polymer concentration is high enough, the helical structure forms a three-dimensional network [2]. As the helical structure is not water soluble, the 3D network structure does not swell in water. Therefore, a large quantity of water is held within the network structure, affording an agarose gel.

We have investigated the solvent transport behaviour of constrained agarose gel as a model system for studying flavour release behavior [3,4]. Compression load occurs when the agarose gel is compressed, but the load is relaxed. We found that the volume decreased with time by approximately the same amount as the compression load relaxation time when the change in gel volume was observed simultaneously. This phenomenon can be successfully explained using stress-diffusion coupling theory [5], which shows that agarose gels can be treated as soft porous bodies.

As most food flavourings are water soluble, solvent transport behaviour and flavour release are expected to be closely related. Agarose is often applied as a gelling agent in sweet desserts. Therefore, changes in the physical properties of agarose gel with the addition of saccharide (sweetener) are an important research theme in food processing. The mechanical and thermal properties of agarose reportedly improve with sucrose addition [6,7]. Furthermore, the syneresis often observed in gelatinous food containing agarose is controlled by sucrose addition [8]. A comprehensive review of sucrose release behaviour from agarose gels has been published [9]. However, the most of these studies are on the passive diffusion behaviour of sucrose in agarose gels. We can investigate the squeeze-out speed of the solvent from the compressed agarose gels using our unique experimental system [3]. It is worth considering flavour release during the eating action.

2. Results and Discussion

2.1. Relationship between Compression Load Relaxation and Volume Change

The time profile of the change in the compression load of agarose gel (1.5 wt %) is shown in Figure 1. The compression load decreased exponentially with time, but did not relax completely, reaching a plateau after 10^4 s. We analysed this relaxation behaviour using a stretched exponential function (Equation (1)):

$$L(t) = L_0 \exp\left(\left(-\frac{t}{\tau_M}\right)^a\right) + L_r \tag{1}$$

where L_0 is the relaxed component of the compression load, τ_M is the relaxation time, L_r is the residue of the compression load, and a is the stretched index. The solid line in Figure 1 represents the fitting result using Equation (1) (the actual values of the parameters are listed in Table 1 as "control"). The experimental data fitted well with Equation (1), indicating that the compression load of agarose gels relaxed over time, but not completely.

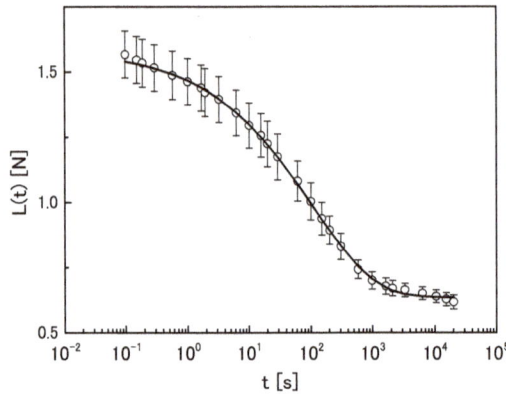

Figure 1. Time profile of compression load for 1.5-wt % agarose gel. The line represents the best fit result of analysis using Equation (1).

The time profile of the volume change in agarose gel (1.5 wt %) is shown in Figure 2. The solid line in Figure 2 is the result of analysis using Equation (2) (the actual values of the parameters are listed in Table 2 as "control")

$$V(t) = V_0 \exp\left(\left(-\frac{t}{\tau_V}\right)^b\right) + V_r \tag{2}$$

where V_0 is the reduced component of the volume change ration, τ_V is the time constant of the volume change speed, V_r is the residue of the volume change ratio, and b is the stretched index. Comparing Figure 1 with Figure 2 showed that this decreasing behaviour was synchronous. This was not coincidental, but attributed to a correlation between these two properties. As we have reported previously [3,4], this correlation can be explained using the stress-diffusion coupling theory [5]. By assuming that agarose gels are soft porous bodies, solvent in the capillary would be squeezed out by compression. Therefore, the speed of this solvent removal from the constrained gels was estimated from the mechanical relaxation. This was valuable because measuring the compression load was easier than measuring the solvent transport speed.

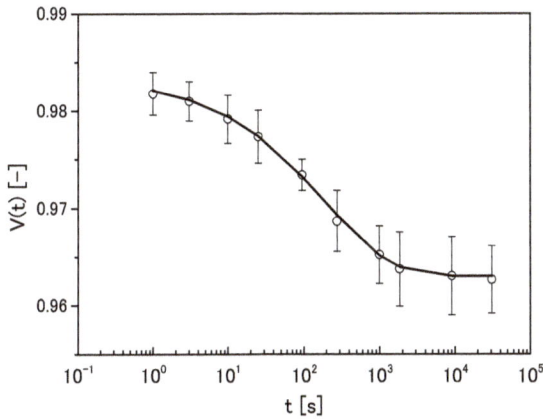

Figure 2. Time profile of the volume change in 1.5-wt % agarose gel. The line represents the best fit result of analysis using Equation (2).

2.2. Effect of Sucrose and Xylitol on Solvent Transport from Constrained Agarose Gels

To determine the effect of sucrose and xylitol on the solvent transport speed, agarose gels (1.5 wt %) were prepared with solvent containing sucrose or xylitol (10–50 wt %), denoted herein as S-Y and X-Y, respectively, where Y indicates the amount of sucrose or xylitol in the solvent (in wt %).

The compression load change and volume change in samples containing sucrose are shown in Figure 3a,b, respectively. The compression load increased with increasing sucrose concentration, as clearly shown in Figure 3a. Furthermore, the relaxation time appeared to increase with increasing sucrose concentration.

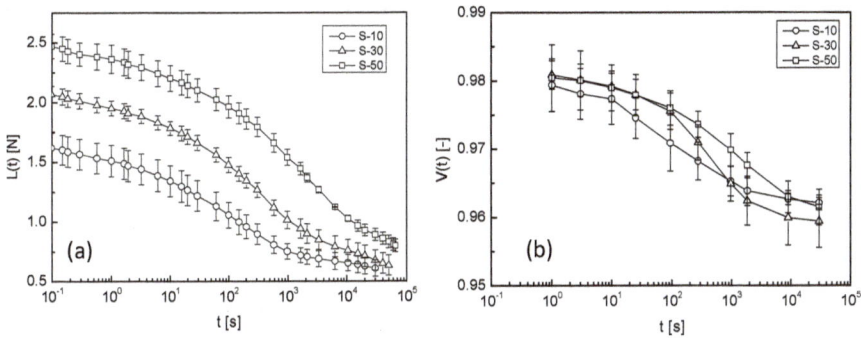

Figure 3. Time profile of (**a**) compression load and (**b**) volume change in samples containing sucrose. Circles, triangles, and squares denote S-10, S-30, and S-50 samples, respectively. Error bars represent the standard deviation. The lines are an eye-guide.

The results obtained for samples containing xylitol are shown in Figure 4. Xylitol addition had similar effects on $L(t)$ and $V(t)$ to sucrose addition, namely, increased $L(t)$ and increased time taken for the volume change with increasing sweetener concentration.

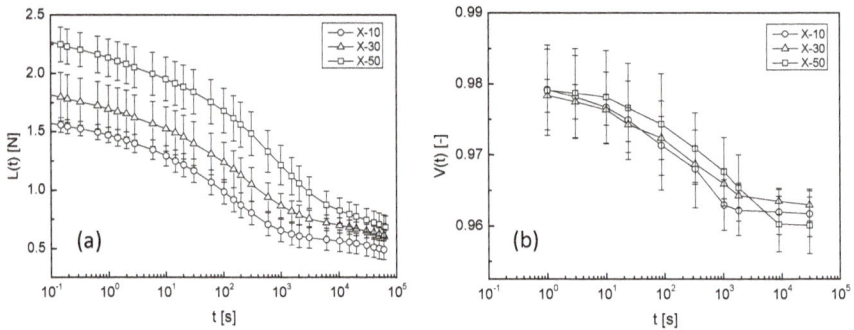

Figure 4. Time profile of (**a**) compression load and (**b**) volume change in samples containing xylitol. Circles, triangles, and squares denote X-10, X-30, and X-50 samples, respectively. Error bars represent the standard deviation. The lines are an eye-guide.

For a quantitative evaluation of the effect of sucrose and xylitol, the experimental curves ($L(t)$ and $V(t)$) were analysed using Equations (1) and (2) for the entire samples. The results are listed in Tables 1 and 2.

Table 1. Characteristic parameters of compression load relaxation analysed using Equation (1).

Samples		L_0 (N)	L_r (N)	τ_M (s)	a (-)
control		1.01 ± 0.119	0.602 ± 0.0572	114 ± 33.1	0.412 ± 0.0220
Sucrose	S-10	1.07 ± 0.108	0.627 ± 0.0752	128.9 ± 26.5	0.355 ± 0.0552
	S-20	1.20 ± 0.0969	0.708 ± 0.0663	2235 ± 46.9	0.377 ± 0.0311
	S-30	1.42 ± 0.0694	0.679 ± 0.100	385 ± 80.7	0.390 ± 0.0378
	S-40	1.58 ± 0.0970	0.695 ± 0.0475	782 ± 170	0.362 ± 0.0302
	S-50	1.77 ± 0.191	0.765 ± 0.141	1970 ± 648	0.367 ± 0.0610
Xylitol	X-10	1.05 ± 0.0832	0.559 ± 0.107	143 ± 41.9	0.422 ± 0.0187
	X-20	1.12 ± 0.0961	0.620 ± 0.0260	163 ± 36.6	0.361 ± 0.0353
	X-30	1.12 ± 0.149	0.669 ± 0.0591	227 ± 78.7	0.360 ± 0.0469
	X-40	1.31 ± 0.0937	0.795 ± 0.0843	407 ± 91.6	0.388 ± 0.0296
	X-50	1.67 ± 0.105	0.680 ± 0.105	749 ± 257	0.340 ± 0.0386

Table 2. Characteristic parameters of volume change analysed using Equation (2).

Samples		$V_0 \times 10^2$ (-)	$V_r \times 10^2$ (-)	τ_V (s)	b (-)
control		2.06 ± 0.221	95.5 ± 1.70	202 ± 72.0	0.514 ± 0.0819
Sucrose	S-10	2.07 ± 0.144	96.2 ± 0.131	233 ± 96.4	0.423 ± 0.0255
	S-20	2.13 ± 0.348	96.0 ± 0.273	390 ± 191	0.518 ± 0.142
	S-30	2.15 ± 0.309	96.0 ± 0.403	597 ± 109	0.629 ± 0.0649
	S-40	1.95 ± 0.201	95.9 ± 0.264	1340 ± 236	0.538 ± 0.0481
	S-50	1.69 ± 0.579	96.1 ± 0.300	1400 ± 806	0.708 ± 0.131
Xylitol	X-10	1.97 ± 0.0671	95.8 ± 0.600	283 ± 108	0.625 ± 0.159
	X-20	1.76 ± 0.111	95.8 ± 0.241	300 ± 52.1	0.592 ± 0.0583
	X-30	1.60 ± 0.149	95.7 ± 0.248	365 ± 201	0.565 ± 0.144
	X-40	1.93 ± 0.135	96.2 ± 0.287	500 ± 82.6	0.552 ± 0.0624
	X-50	1.98 ± 0.237	96.0 ± 0.410	1130 ± 104	0.507 ± 0.0500

To easily check the effect of adding sweeteners on the solvent transport behaviour, the mechanical relaxation time (τ_M) and the time constant (τ_V) of the volume change of the agarose gel containing the sweeteners are plotted against their concentrations. The effect of sucrose and xylitol addition are shown in Figures 5 and 6, respectively. Sucrose and xylitol were found to effectively reduce the speed of solvent transport from the mechanically-constrained gel. However the solvent transport speed reducing effect of xylitol is less than sucrose. A possible explanation for this phenomena provided below (in Section 2.4).

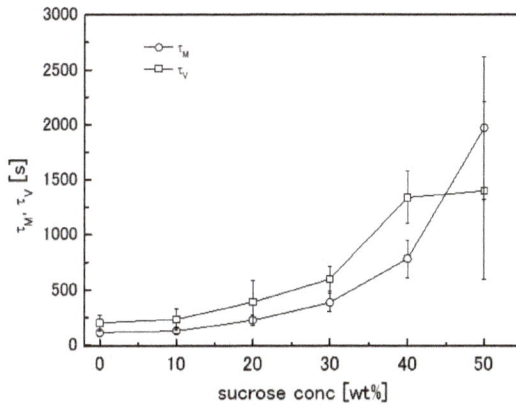

Figure 5. Dependence of time constants τ_M (circles) and τ_V (squares) on sucrose concentration. Error bars represent the standard deviation.

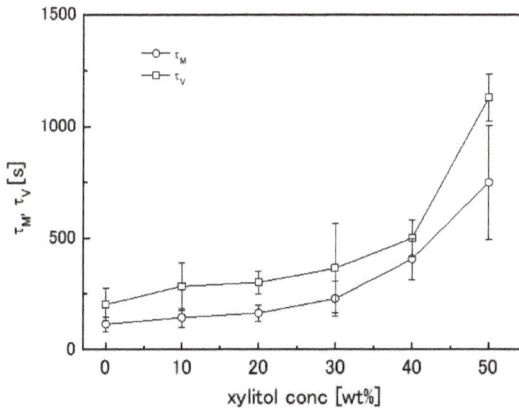

Figure 6. Dependence of time constants τ_M (circles) and τ_V (squares) on xylitol concentration. Error bars represent the standard deviation.

2.3. Influence of Sweetener Addition on Network Structure of Agarose Gel

The network structure of agarose gel has been reported to change with sucrose addition [6–8]. The enthalpy of gelation and Young's modulus reportedly increased with sucrose addition. In this study, the compression load also increased with sucrose addition, as shown in Figure 3a. In contrast, the influence of xylitol on the Young's modulus of agarose gel has rarely been reported. Xylitol was also clearly effective in increasing the Young's modulus of agarose gel, as shown in Figure 4a. L_0 in Equation (1) represents the relaxed component of the compression load. The sweetener concentration-dependency of L_0 is shown in Figure 7. L_0 clearly increased with increasing sweetener concentration. This result suggested that a change occurred in the network structure formed by agarose gel with sucrose or xylitol addition. Therefore, freeze-dried gel samples were observed using SEM.

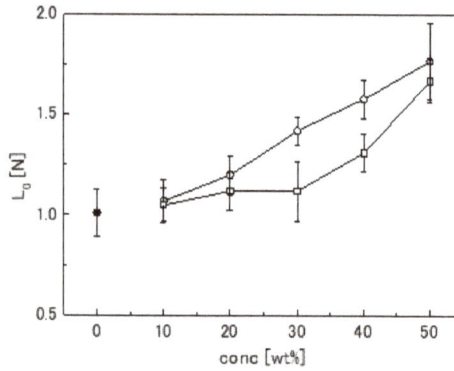

Figure 7. Gel strength (L_0 in Equation (1)) of samples containing various sweetener concentrations. Closed circles, open circles, and open squares denote the control (1.5-wt % agarose), sucrose series, and xylitol series, respectively.

Scanning Electro-Microscopy (SEM) images of the agarose gel structure with sucrose addition are shown in Figure 8. As the sucrose concentration increased, the mesh size of the agarose gel network seemed to become smaller. The results for xylitol addition using a similar method are shown in Figure 9. A decrease in the mesh size of the agarose gel network was observed with xylitol addition, as observed for sucrose addition above. Agarose molecules, with their spiral structure, are known to associate to form a gel network structure. The sequence state of the spiral structure has been reported to change with sucrose addition, resulting in the strength of the gel network structure changing. This phenomenon was thought to cause the increasing Young's modulus observed in this study. Xylitol also clearly had a similar effect on the agarose gels in this study.

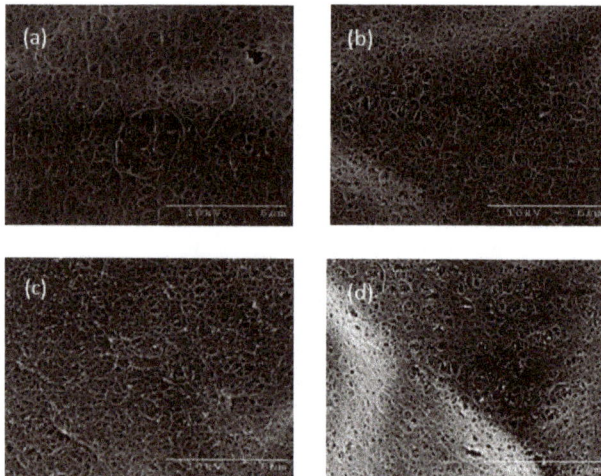

Figure 8. SEM images of (**a**) S-10, (**b**) S-20, (**c**) S-30, and (**d**) S-50. The scale bars indicate 5 μm.

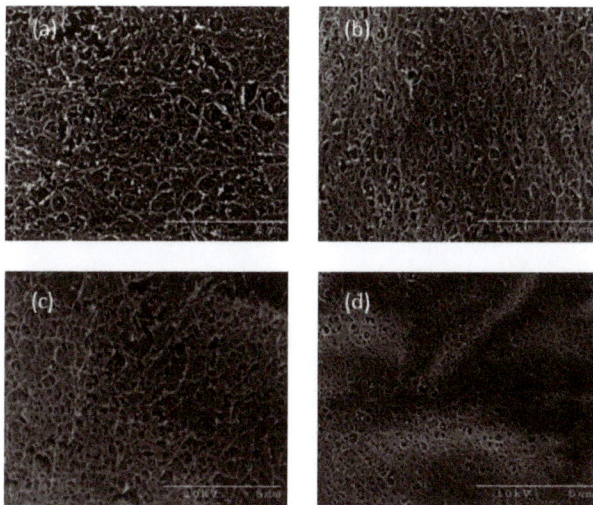

Figure 9. SEM images of (**a**) X-10, (**b**) X-20, (**c**) X-30, and (**d**) X-50. The scale bars indicate 5 μm.

2.4. Relationship between Structural Changes and Solvent Transport Behaviour of Agarose Gel Networks

The syneresis of agarose gel is thought to be empirically controlled by sucrose addition. However, the mechanism of this phenomenon on the molecular level remains unclear. The improvements in the mechanical characteristics resulting from sucrose addition discussed in Section 2.3 are partly due to changes in the gel network. The moving speed of solvent molecules is thought to decrease intuitively owing to the mesh size of the gel network decreasing. However, the mesh size of the gel network, as observed by SEM (Figures 8 and 9), was too large to prevent solvent transport. Furthermore, their size was estimated to range from 10 to 100 nm [10–12].

As mentioned above, the solvent transport behaviour of mechanically-constrained agarose gel can be explained using the stress-diffusion coupling theory, which treats agarose gels as soft porous bodies. Considering the agarose gels to be soft porous bodies, an increase in inner pressure would occur during compression, which should squeeze solvent out of capillaries in the gel. The flow quantity (Q) can be estimated using the Hagen–Poiseuille law (Equation (3)):

$$Q \propto \frac{r^4}{\eta L} P \tag{3}$$

where P is the increase in pressure, r is the capillary radius, L is the capillary length, and η is the flow liquid viscosity. The solvent viscosity was the factor determining the flow quantity if the shape of the capillary and pressure increase were constant. According to the literature, sucrose–water solutions have viscosities of 1.06 mPa s (10 wt %) and 10.0 mPa s (50 wt %) at 30 °C. Using these viscosity values and Equation (3), the flow quantity of S-10 was estimated to be ten-fold that of S50. In contrast, the value of τ_M for S-10 was 15-fold higher than that of S-50, with the τ_V for S-10 six-fold higher than that of S-10 (see Table 1). These values were similar to the estimated value mentioned above. Therefore, the reduction in solvent transport speed resulting from sucrose addition might be mainly due to the rise in solvent viscosity. The effect of xylitol was expected to be similar to that of sucrose. As viscosity data for xylitol–water solutions were not available in the literature, the viscosities were measured, giving approximately 1 mPa s and 6 mPa s for 10-wt % and 50-wt % solutions, respectively. The viscosity thickening effect of xylitol was slightly inferior to that of sucrose, but in the same order of magnitude, and the increase in τ_M and τ_V was thought to be dependent on the solute levels, with similar dependencies observed in both the xylitol and sucrose systems. Water molecules reportedly strongly hydrate sugar molecules in agarose, water, and sugar-containing systems, resulting in their removal from the vicinity of the agarose molecules. Therefore, sucrose and xylitol may become concentrated in the low density phase of the agarose gel, namely in capillary of soft porous body. If the sweetener aqueous solution is concentrated locally, the viscosity of the solvent rises, and is thought to be observed as a drop in solvent transport speed from the mechanically-constrained agarose gel macroscopically.

3. Conclusions

The effect of sweetener on the speed of solvent transport speed from mechanically-constrained agarose gels was investigated. Empirically, it was known that sucrose prevents syneresis from agarose gels, but the mechanism of this phenomenon remained unclear. Herein, we estimated the solvent transport speed from constrained agarose gels containing sweeteners using a unique experimental system. Both sucrose and xylitol caused a concentration-dependent reduction in the solvent transport speed. This phenomenon is attributed to two factors: (i) Changes in the microscopic structure of the agarose gels caused by adding sweeteners, and (ii) changes in the viscosity of the solvent containing sweeteners. We showed that agarose gels can be treated as soft porous bodies in a previous study. As the viscosity of the solvent would increase with the addition of sweetener, it is expected that the flow quality (speed) would decrease according to the Hagen–Poiseuille law.

4. Materials and Methods

4.1. Materials

Agarose type-IV (Sigma-Aldrich, St. Louis, MO, USA) containing less than 0.25% sulfate was purchased and used without further purification. For all samples, the agarose concentration was 1.5 wt %. Sucrose and xylitol aqueous solution were used as solvents for the gels. Their concentrations were ranging from 0 to 50 wt % in distilled water. Sample gels were prepared as follows; Agarose powder was dispersed in the solvent and stirring at room temperature for 18 h. The dispersion was heated at 95 °C for 1 h to completely dissolve. The hot solution was poured into a plastic tube (ϕ = 20 mm) and both ends of the tube were sealed. The tube was then immediately placed in a water bath for temperature control. After 24 h of quenching in the water bath at 10 °C, the sample gels were cut to a length of about 20 mm using a razor blade. The cylindrical gels were immersed in solvent, which was used in the preparation of gels, and incubated at 5 °C for at least five days to reach an equilibrated state before the experiments. The diameter and length of gel samples were measured with a caliper immediately before experiments.

4.2. Monitoring of Compression Load and Volume Change

The measurement system has been described elsewhere [3]. Compression and monitoring of the compression load were performed using an INSTRON MINI 55 instrument (Instron, Norwood, MA, USA). To cancel the surface tension at the surface of gel samples and prevent drying, the gel samples were immersed in their solvent during experiment and kept their temperature at 30 °C. The sample gel was compressed at a rate of 1 mm/s and, when the compression strain reached 0.05, the compression was kept constant. Changes in the compression load were monitored for 18 h. The pictures of the side view of the sample was taken using a charge-coupled device (CCD) camera. The top and bottom surfaces of the cylindrical gel were sealed with cyanoacrylate to prevent slipping. As the gel deformed to a barrel-like shape, the width was measured at five different points and the average width (w_m) was obtained. The height of the gel was determined during compression load measurements and the gel volume (v) was calculated using Equation (4), with each measurement performed at least in triplicate.

Digital images of the samples were analysed using image analysis software (Image-J (NIH), Bethesda, MD, USA).

$$v = \pi \left(\frac{w_m}{2} \right)^2 \cdot h \tag{4}$$

4.3. Viscosity Measurements

The apparent viscosities of the solvents were measured by a rotation rheometer (ARES: TA instruments, New Castle, DE, USA) equipped with a bob (diameter, 16 mm) and cup (diameter, 16.5 mm). The steady state viscosity at 50 s^{-1} was measured at 30 °C.

4.4. Scanning Electron Microscopy (SEM) Observation

The microstructure of the agarose gels was observed using a scanning electron microscope (HITACHI S-2460N, Tokyo, Japan). The sample gels were immersed in distilled water for at least 10 days to remove sucrose or xylitol. The washed gels were rapidly cooled in liquid nitrogen, and then the completely freeze-dried samples were cleaved to expose the torn surfaces and coated with Pt-C on the SEM sample table.

Acknowledgments: This work was supported by JSPS KAKENHI Grant Number 25410231.

Conflicts of Interest: The authors declare no conflict of interest

References

1. Morris, V.J. Gelation of polysaccharides. In *Functional Properties of Food Macromolecules*, 2nd ed.; Hill, S.E., Ledward, D.A., Mitchell, J.R., Eds.; An Aspen Publication: Silver Spring, MD, USA, 1998; pp. 143–168.
2. Arnott, S.; Fulmer, A.; Scott, W.E.; Dea, I.C.M.; Moorhouse, R.; Rees, D.A. The agarose double helix and its function in agarose gel structure. *Mol. Biol.* **1974**, *90*, 273–284. [CrossRef]
3. Kaneda, I.; Iwasaki, S. Solvent transportation behaviour of mechanically constrained agarose gels. *Rheol. Acta* **2015**, *54*, 437–443. [CrossRef]
4. Kaneda, I.; Sakurai, Y. Effect of glycerol on solvent transportation behaviour of mechanically constrained agarose gels. *Food Hydrocoll.* **2016**, *61*, 148–154. [CrossRef]
5. Yamane, T.; Doi, M. The stress diffusion coupling in the swelling dynamics of cylindrical gels. *J. Chem. Phys.* **2005**, *122*, 84703. [CrossRef] [PubMed]
6. Watase, M.; Nishinari, K.; Williams, P.A.; Phillips, G.O. Agarose gels: Effect of sucrose, Glucose. Urea and Guanidine Hydrochloride on the Rheological and Thermal Properties. *J. Agric. Food Chem.* **1990**, *38*, 1181–1187. [CrossRef]
7. Normand, V.; Aymard, P.; Lootens, D.; Amici, E.; Plucknett, K.P.; Frith, W.J. Effect of sucrose on agarose gels mechanical behaviour. *Carbohydr. Polym.* **2003**, *54*, 83–95. [CrossRef]
8. Wang, Z.; Yang, K.; Brenner, T.; Kikuzaki, H.; Nishinari, K. The influence of agar gel texture on sucrose release. *Food Hydrocoll.* **2014**, *36*, 196–203. [CrossRef]
9. Nishinari, K.; Fang, Y. Sucrose release from polysaccharide gels. *Food Funct.* **2016**, *7*, 2130–2146. [CrossRef] [PubMed]
10. Boral, S.; Bohidar, H.B. Hierarchical structure in agarose hydrogels. *Polymer* **2009**, *50*, 5585–5588. [CrossRef]
11. Dai, B.; Matsukawa, S. Elucidation of gelation mechanism and molecular interactions of agarose in solution by 1H NMR. *Carbohydr. Res.* **2013**, *365*, 38–45. [CrossRef] [PubMed]
12. Xiong, J.-X.; Narayanan, J.; Liu, X.-J.; Chong, T.K.; Chen, S.B.; Chung, T.S. Topology evolution and gelation mechanism of agarose gel. *J. Phys. Chem. B* **2005**, *109*, 5638–5643. [CrossRef] [PubMed]

gels

MDPI

Article

Dynamics of Spinodal Decomposition in a Ternary Gelling System

Yutaro Yamashita, Miho Yanagisawa [†] and Masayuki Tokita *

Department of Physics, Graduate School of Science, Kyushu University, 744 Motooka, Fukuoka 810-0935, Japan; yama_cf@yahoo.co.jp (Y.Y.); myanagi@cc.tuat.ac.jp (M.Y.)
* Correspondence: tokita@phys.kyushu-u.ac.jp; Tel.: +81-92-802-4095
† Current address: Department of Applied Physics, Faculty of Engineering, Tokyo University of Agriculture and Technology, 2-24-16 Naka-cho, Koganei, Tokyo 184-8588, Japan.

Received: 26 February 2018; Accepted: 19 March 2018; Published: 22 March 2018

Abstract: The phase diagram and phase transitions of the ternary system of gelatin, water and poly(ethylene glycol) oligomers were studied as a function of the weight fraction of gelatin and the weight fraction and molecular weight of poly(ethylene glycol) oligomers. It was found that both phase separation and the sol-gel transition occur in this ternary system. The relative position of the phase separation line and the sol-gel transition line depends on the weight fraction and the molecular weight of the poly(ethylene glycol) oligomer that coexists in the solution. All aspects of the phase diagram are sensitive to the molecular weight of the poly(ethylene glycol) oligomer. Since the phase separation line crosses the sol-gel transition line in the phase space that is created by the temperature and the weight fraction of gelatin, the phase space is typically divided into four regions, where each region corresponds to a definite phase. The transitions between mutual phases were studied using the light-scattering technique.

Keywords: sol-gel transition; site-bond correlated-percolation model for polymer gelation; gelation temperature; cloud point temperature; spinodal temperature; spinodal decomposition

1. Introduction

There are many water-soluble polymers distributed throughout the natural world. Some of these can be synthesized artificially, while others solely originate from biological systems. Since water-soluble polymers are environmentally sustainable, they are widely used in science, technology and even in our day-to-day lives. The phase behavior and the phase transition of polymer solution play important roles in the fabrication processes of polymeric materials. Therefore, it may be important to obtain information on the phase structure of the water-soluble polymer in the solution state for the effective fabrication of many practical commodities. In this paper, we report on the phase behaviors of gelatin, water and poly(ethylene glycol) (PEG) in a system.

Gelatin is a typical protein derived from collagen in the skin and bones of animals. It is well known that the aqueous solution of gelatin shows sol-gel transition. The gel of gelatin is widely used in food, and hence, the sol-gel transition of gelatin solution has been studied in detail thus far [1–3]. On the other hand, PEG is also well known as a neutral water-soluble polymer. The polymer and its oligomers interact with proteins, and hence, they are used as precipitating agents of protein in protein crystallography. The phase diagram of the aqueous PEG system has been also studied [4]. It is clarified that the aqueous PEG solution shows the phase separation only at very higher temperature regions rather than the room temperature region.

It has been reported that the phase separation becomes obvious when methanol is added to the aqueous solution of gelatin as the third component of the system [5]. Although the results are not systematic, it was also reported that the aqueous gelatin solution shows phase separation when a trace amount of large-molecular-weight PEG is added to the solution [6]. Therefore, the ternary system of the aqueous gelatin with PEG has attracted much attention since these studies, and then, many more studies have been done, especially in pharmaceutical science [7–11]. However, the systematic study of the phase behaviors of the ternary system of gelatin, water and PEG has yet to be done.

Here, we report on the systematic study of the phase behaviors in the ternary system of gelatin, water and PEG. A typical phase diagram of the system has been described in the previous report [12]. The key point of the previous study is the choice of PEG. We choose the oligomers of PEG with various molecular weights in this study. As such, the weight fraction of PEG oligomer can be changed widely, and hence, the overall aspects of the phase behaviors of the ternary system of gelatin, water and PEG were revealed. In this study, the concentration range of PEG in the system was changed widely to obtain the phase diagram of the system in the entire parameter space. The phase separation of the solution was observed in addition to the sol-gel transition of the solution in the ternary system. It was found that two phase transition lines of the phase separation and the sol-gel transition cross in the $T - \varphi_g$ phase space, where T and φ_g represent the temperature and the weight fraction of gelatin, respectively. Therefore, the phase space of the ternary system is typically divided into four regions corresponding to the definite phases. The phase transitions were investigated mainly by the microscopic observations in the previous study. Thus, the light-scattering measurements were taken to clarify the characteristics of the phase transitions between these four phases in this study.

2. Results

2.1. Phase Diagram

In Figure 1, we show the phase diagrams of the system that are determined under various conditions of the weight fraction of gelatin, as well as the weight fraction and the molecular weight of the oligomer PEG. The cloud point and the gelation temperature are plotted as a function of the weight fraction of gelatin in each phase diagram. The phase diagram itself is arranged by two parameters of the weight fraction and the molecular weight of the oligomer PEG.

It is clear from the phase diagram that the aqueous gelatin solution only shows the sol-gel transition, as can be seen in the bottom left corner. The phase diagram of the aqueous gelatin solution is similar to those seen in previous reports [2,5]. In the phase diagrams of diethylene glycol (PEG-100) systems, it was found that the solution only shows sol-gel transition when the weight fraction of PEG-100 is 0.1. Although data are not shown, the same phase diagrams were obtained in the systems of triethylene glycol (PEG-150) and tetraethylene glycol (PEG-200) at a weight fraction of 0.1 [12]. In contrast, the phase separation became obvious when the weight fraction of PEG-100 was increased to 0.4. The results indicate that the solution of PEG-100 is a poor solvent for gelatin when the weight fraction is increased to 0.4 because the phase separation of the polymer solution is a characteristic behavior that occurs in poor solvent systems [13].

It was also found from the series of the phase diagrams with the weight fraction of PEG in the 0.1 system that the phase separation line appears when the molecular weight of the PEG oligomer is increased at a constant weight fraction. In addition to this, the critical temperature is raised and is shifted to the lower concentration region of gelatin with the molecular weight of the PEG oligomer that coexists in the system. The phase diagrams shown in Figure 1 indicate that the relative position of the phase separation line and the sol-gel line varies with the weight fraction and the molecular weight of the PEG oligomer. It was found that the phase diagrams of PEG-100 at a weight fraction of 0.4 and that of PEG-400 at a weight fraction of 0.1 in the system become similar. A similarity in the phase diagrams was also observed between PEG-400 at a weight fraction of 0.2 and PEG-1000 at a weight fraction of 0.08 in the system. The results indicate that the phase diagram of the ternary system of gelatin, water and PEG can be controlled by the appropriate choice of the molecular weight and the weight fraction of PEG oligomer.

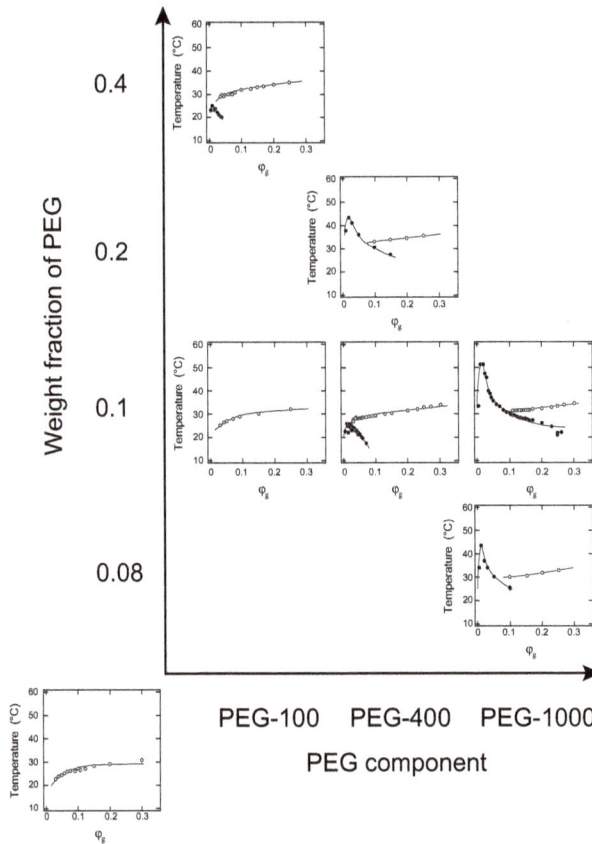

Figure 1. The phase diagrams of systems with respect to gelatin, water and oligomers of poly(ethylene glycol) (PEG). The cloud point (closed circles) and the gelation temperature (open circles) are plotted as a function of the weight fraction of gelatin, φ_g. Each phase diagram is aligned as a function of the weight fraction of PEG and the molecular weight of PEG. The phase diagram of the aqueous gelatin solution is shown in the bottom left corner. The series of phase diagrams at a weight fraction of 0.1 is a reproduction from a previous report [12]. PEG-100: diethylene glycol; PEG-400: PEG at M. W. ~ 400; PEG-1000: PEG at M. W. ~ 1000.

Among others, the series of the phase diagrams at a PEG weight fraction of 0.1 clearly shows the effects of the molecular weight on the phase diagram. Phase diagrams of the system at a weight fraction of PEG oligomers of 0.1 (PEG-100, PEG-400, PEG-1000 systems) in Figure 1 are summarized as Figure 2 with the phase diagram of the aqueous gelatin solution, where phase separation temperatures and sol-gel transition temperatures are shown in closed circles and colored circles, respectively. Only sol-gel transition is observed in the system of PEG-100 at a weight fraction of 0.1, while both the sol-gel transition and the phase separation are observed in the systems of PEG-400 (M. W. ~ 400) and PEG-1000 (M. W. ~ 1000) at the same weight fraction of PEG. The appearance of the phase diagram is completely distinct from that of the PEG-100 system. Since the weight fraction of the oligomer in the solution is the same, the number of oligomers in the solution of PEG-1000 is only about one-tenth of that of the PEG-100 system. The results, therefore, indicate that the molecular weight of the PEG oligomer is very effective at changing the phase separation line rather than changing the sol-gel transition line. As the PEG chain becomes longer, the phase separation temperature increases dramatically. In contrast, the positions of the sol-gel transition lines are almost independent of the molecular weight of PEG under the present experimental conditions.

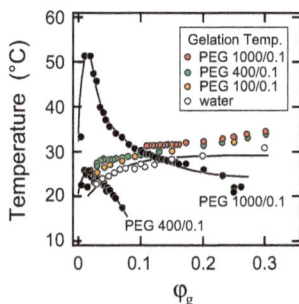

Figure 2. The phase diagrams in the series of PEG oligomers at a weight fraction of 0.1. Sol-gel transition temperatures are expressed by colored circles, and the phase separation temperatures are expressed by solid (black) circles, respectively. The open circles (white) represent the sol-gel transition line of the aqueous gelatin solution.

2.2. Light Scattering

When the phase structure of the system was revealed, we were interested in how the transition between phases occurred. In the phase diagram in Figure 1, the systems of PEG-100 at a weight fraction of 0.4 and PEG-400 at a weight fraction of 0.1 are too delicate to study the transition behaviors because the sol-gel transition line terminates near the critical point of the phase separation line. Such a point is physically interesting to study, but difficult to analyze, and hence, we have left it for future studies. Accordingly, we choose the system of PEG-1000 at a weight fraction of 0.1 to obtain the fundamental characteristics of the phase transition between phases. Since two phase boundaries cross in this system, the $T - \varphi_g$ phase space is divided into four regions, I, II, III and IV, as shown in Figure 3. Each region corresponds to the definite phase as follows.

- Region I ⟶ one-phase sol
- Region II ⟶ one-phase gel (transparent)
- Region III ⟶ two-phase sol
- Region IV ⟶ two-phase gel (opaque)

Among the above regions, Regions I and II correspond to the simple sol and the simple gel phases so far observed in the aqueous gelatin solution. The difference between the gel formed in Regions II and IV is clear in appearance, but the definition of the phase separation line in the gel state is not clear enough. Finally, the presence of the phase boundary between III and IV is not clarified in the present study.

The light-scattering measurements are made under the following quench conditions, and the results are given in Figure 4.

1. $\varphi_g = 0.05$, T is changed from 40.0 to 10.0 °C by single quench (Region I → Region III).
2. $\varphi_g = 0.1$, T is changed from 38.0 to 15.0 °C by single quench (Region I → Region IV).
3. $\varphi_g = 0.2$, T is changed from 60.0 to 15.0 °C by single quench (Region I → (Region II) → Region IV).
4. $\varphi_g = 0.2$, T is changed from 60.0 to 30.0, then to 15.0 °C by double quench (Region I → Region II → Region IV).

Since the scattering profiles shown in Figure 4 are strongly overlapped, the scattering profiles are shifted vertically and shown in Figure 5 for the sake of clarity.

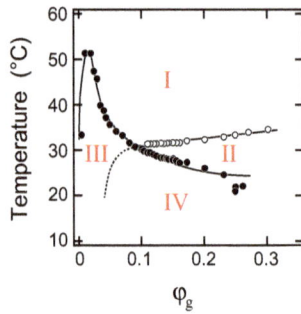

Figure 3. The phase diagram of the PEG-1000 system at the weight fraction of 0.1 in Figure 1. The $T - \varphi_g$ phase space is divided into four regions. Region I: one-phase sol; Region II: one-phase gel; Region III: two-phase sol; Region IV: two-phase gel. The presence of the phase boundary between two phases of III and IV is not obvious in the present study.

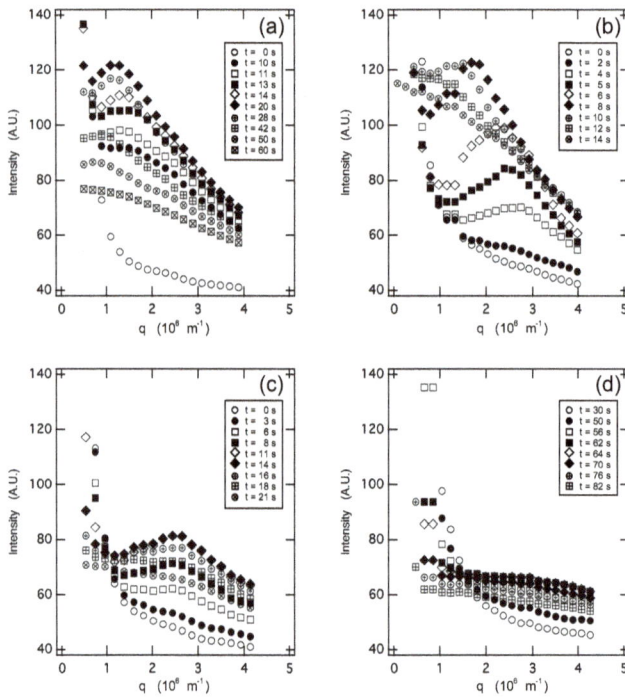

Figure 4. The light-scattering profile from the system undergoing phase transition. The relative intensity of the scattered light from the sample is plotted as a function of the scattering vector, q. The series of the scattering profiles is plotted as a function of the elapsed time after the temperature change, which is given in each figure. The systems are $\varphi_g = 0.05$ (T; 40.0 → 10.0 °C) (**a**); $\varphi_g = 0.1$ (T; 38.0 → 15.0 °C) (**b**); $\varphi_g = 0.2$ (T; 60.0 → 15.0 °C, single quench) (**c**); and $\varphi_g = 0.2$ (T; 30.0 → 15.0 °C) (second quench of the double quench) (**d**). The weight fraction of PEG-1000 is fixed at 0.1.

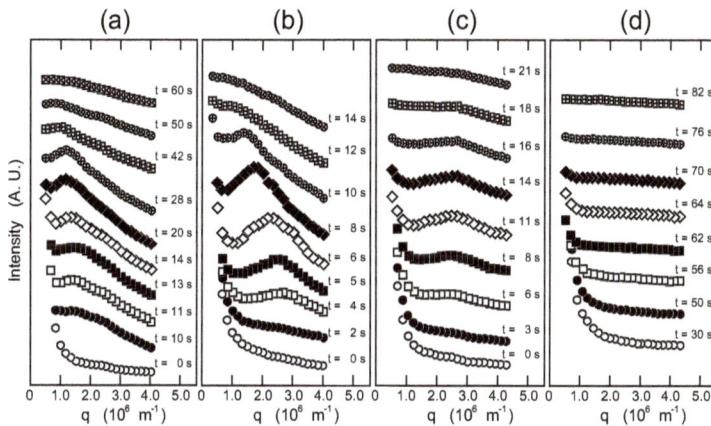

Figure 5. The scattering profiles shown in Figure 4 are vertically shifted with respect to each other to show the time evolution of the maximum that appeared in the scattering profile. Symbols are the same as used in Figure 4. The systems are $\varphi_g = 0.05$ (T; $40.0 \rightarrow 10.0$ °C) (**a**); $\varphi_g = 0.1$ (T; $38.0 \rightarrow 15.0$ °C) (**b**); $\varphi_g = 0.2$ (T; $60.0 \rightarrow 15.0$ °C, single quench) (**c**); and $\varphi_g = 0.2$ (T; $30.0 \rightarrow 15.0$ °C) (second quench of the double quench) (**d**). The weight fraction of PEG-1000 is fixed at 0.1.

2.2.1. System $\varphi_g = 0.05$

It is expected from Figure 3 that the transition that occurs during this process is simple phase separation from the one-phase sol (Region I) to the two-phase sol (Region III). The sol-gel transition of the system may not have significant effects on the phase separation since the phase separation occurs in a rather higher temperature region of about 35 °C. The in situ optical microscope observation indicates that the phase separation proceeds by the nucleation-growth process in the slow cooling process of the solution [12]. In contrast, the light-scattering results from the quenched solution, Figures 4a and 5a, show the characteristic behaviors of the spinodal decomposition. Namely, the scattering intensity shows a maximum after 10 s from the quench of the solution. The magnitude of the scattering vector at which the scattering intensity shows the maximum, q_{max}, is about $q_{max} \sim 2 \times 10^6\,\text{m}^{-1}$. The position of the maximum in the scattering profile moves towards the smaller values of q with time, and then, it becomes invisible to the scattering profile after 60 s from the quench of the solution. The results strongly suggest the emergence and divergence of the density fluctuations. These results are similar to the previous studies that were made in different systems [14–17]. It was, therefore, found that the phase separation of the solution occurs by spinodal decomposition when the solution is quenched from Region I to III. Taking into account the results of the previous in situ optical microscope observation, the phase transition observed here is the simple phase separation.

2.2.2. System $\varphi_g = 0.1$

In this system, the light-scattering results, Figures 4b and 5b, showed the similar behaviors with the system $\varphi_g = 0.05$. However, it is clear from Figures 4b and 5b that the maximum in the scattering profile appears at much earlier time frames than that in the system $\varphi_g = 0.05$: the scattering profile shows a maximum at about $q_{max} \sim 3 \times 10^6\,\text{m}^{-1}$ after 2 s from quenching. The results suggest that the spinodal decomposition is considerably accelerated in the $\varphi_g = 0.1$ system. The quench process of the system is special because the system is cooling through the cross-over point of the sol-gel transition line and the phase separation line. Therefore, both the phase separation and the network formation by the sol-gel transition may equally effect the time evolution of the system.

2.2.3. System $\varphi_g = 0.2$ in Single Quench

The scattering results of this system, Figures 4c and 5c, were totally altered from the previous two systems. The maximum appeared in the scattering profile, but it was observable only in a certain time interval from 6 to 18 s after the quenching of the solution. Besides, the magnitude of the scattering vector at the maximum, $q_{max} \sim 2.5 \times 10^6 \, m^{-1}$, was time independent throughout the observation. The final state of the system is the two-phase gel, which is opaque in appearance. A similar phenomenon was observed in the gelation process of agarose [18].

2.2.4. System $\varphi_g = 0.2$ in Double Quench

The phase transition behavior of this system was studied by the double quench process to reveal the transition behavior of this system in detail. In the double quench process, the solution was firstly quenched into 30.0 °C from the solution state at 60.0 °C where the one-phase gel, which is transparent in appearance, was formed: Region II in Figure 3. The solution was kept at 30.0 °C overnight to ensure the gelation at this temperature. Then, the transparent gel thus obtained was quenched into the phase separation region at 15.0 °C, and the light-scattering profiles were gained. The scattering results, Figures 4d and 5d, did not show any singular behaviors. Only the opaqueness of the gel is increased after the second quench process. The results of the turbidity measurements after the second quench are given in Figure 6 to complement the light-scattering results. It is clear from Figure 6 that the opaqueness of the gel increases immediately after the quenching of the transparent gel from 30.0 to 15.0 °C. The second quench process of the system corresponds to the transition from the homogeneous one-phase gel to the inhomogeneous two-phase gel.

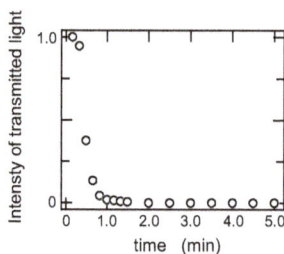

Figure 6. The time evolution of the relative intensity of the transmitted light after the second quench of the system $\varphi_g = 0.2$ from 30.0 to 15.0 °C. The weight fraction of PEG-1000 is 0.1.

3. Discussion

In this study, the interaction between gelatin and PEG was viewed in the form of a phase diagram of the ternary system. It was found from the phase diagrams that the two transitions of the phase separation and the sol-gel transition appear in the $T - \varphi_g$ phase space in the ternary system of gelatin, water and PEG oligomer. The relative position of the phase separation line and the sol-gel transition line was found to depend both on the concentration and the molecular weight of PEG oligomer. The results indicate that the addition of PEG into the aqueous solution of gelatin deteriorates the quality of the solvent as has been observed in the ternary system of gelatin, water and methanol [5]. The phase behaviors of the present ternary system that are shown in Figure 1 are essentially in agreement with the site-bond correlated-percolation theory of the polymer gelation [19,20].

The time evolution of the light-scattering profiles from the solution shown in Figures 4 and 5 indicates that the spinodal decomposition plays essential roles in the quench processes of the system. Therefore, the time evolution of the magnitude of the scattering vector at the maximum that appears in the scattering profile, q_{max}, is shown in Figure 7. The reciprocal of q_{max} is the measure of the correlation length, ξ_{SD}, over which the density fluctuation of the spinodal decomposition correlates. The least squares analysis of the results shown in Figure 7 yields the following.

$$q_{max} \propto t^{-0.4} \quad (\varphi_g = 0.05, 15 \text{ s} \leq t) \tag{1}$$

$$q_{max} \propto t^{-1} \quad (\varphi_g = 0.1, 6 \text{ s} \leq t) \tag{2}$$

$$q_{max} \propto t^{0} \quad (\varphi_g = 0.2, 3 \text{ s} \leq t \leq 18 \text{ s}) \tag{3}$$

The late stage behavior of the $\varphi_g = 0.05$ system is close to the typical behavior of the simple liquid-liquid phase separation, $q_{max} \sim t^{-\alpha}$, where $\alpha = 1/3$ [21]. Since the sol-gel transition temperature is quite similar to that of aqueous gelatin solution without PEG, as shown in Figure 2, the PEG addition might not have affected the gelation of gelatin via triple-helix formation at this concentration of PEG oligomer. In contrast to the dilute system, it is clearly observed that the phase separation is much accelerated in the dense system of $\varphi = 0.1$, where $\alpha \simeq 1$. In this accelerated phase separation, the following two factors are considered: (i) the attractive interaction due to the gelation and (ii) viscoelastic phase separation. Such acceleration of the phase separation has been studied theoretically, as well as experimentally so far [22–27]. Since the sol-gel transition temperature was slightly above 30 °C, entanglement of the polymers might contribute to the elastic nature of the system and bring viscoelastic phase separation. Such viscoelastic phase separation is expected to be observed by adding much longer PEG [6]. Besides, the previous studies suggest that the viscoelastic phase separation is strongly affected by the experimental conditions such as the cooling rate and the shear rate, so detailed and precise experiments are required for the full understanding of the viscoelastic phase separation in this system [28–30]. Such studies are now in progress and will be reported elsewhere.

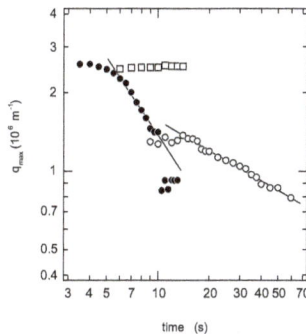

Figure 7. The time evolution of q_{max}. Results are shown for the $\varphi_g = 0.05$ system (open circles), the $\varphi_g = 0.1$ system (closed circles) and the $\varphi_g = 0.2$ system with single quench (open squares). The weight fraction of PEG-1000 is fixed at 0.1. The slopes of the lines drawn for the $\varphi_g = 0.05$ system are -0.4 and for the $\varphi_g = 0.1$ system, -1.

On the other hand, the phase transition in the $\varphi_g = 0.2$ system is completely different from that of the previous two systems as shown in Figures 4, 5 and 7, $q_{max} \propto t^{0}$. It is clear from Figure 3 that the phase separation line is buried in the gel phase in this system. Therefore, the quench depth is deeper for the gelation than that of phase separation, and hence, the gelation proceeds faster. The density fluctuation of the spinodal decomposition appears in the system; however, it cannot grow beyond the size of the polymer network of the gel. Then, the density fluctuations are spontaneously pinned into the polymer network of the gel in the intermediate stage. Finally, the density fluctuations once pinned in the polymer network of the gel are smeared out by the cross-linking reaction. The spontaneous pinning of the density fluctuations was reported in the late stage of the spinodal decomposition in the polymer mixture and the gelation process of agarose solution [18,31]. We, then, carried out the double quench experiment to clarify the competition of the gelation and the phase separation in the $\varphi_g = 0.2$ system. The transparent gel was formed in the first quench process from Region I to II. The double quench process, therefore, corresponds to the case where the gelation reaction proceeds much earlier than the phase separation. It is clear that the time evolution of the light-scattering profile does not show any characteristic behavior, and only the turbidity of the gel increases after the second quench process as shown in Figures 4d, 5d, and 6. Although the polymer network of the gel is formed, the phase separation can occur in the system. This further suggests that the phase separation and the sol-gel transition are independent phenomena under the present experimental conditions. The results are what would be expected from the site-bond correlated-percolation model of the polymer gelation. Since the phase separation occurs in the polymer network of the gel, the phase separation is restricted to the microscopic scale by the polymer network of the gel, which prevents the divergence of the density fluctuations. Such a phase separation has been called microphase separation so far. The microphase separation within the gel is also expected in the equivalent phase diagram of the gelling system [32]. The phase boundary that separates Phases II and IV, which is drawn as the extrapolation of the phase separation line in the lower concentration region in Figure 3, should be recognized as the microphase separation line in the homogeneous gel. The physical meaning of the phase separation line changes at the cross-over point of the sol-gel transition line and the phase separation line.

The time evolution of the scattering light intensity from the system is also discussed together with the time evolution of q_{max} in the dynamics of the phase separation. The scattering light intensity at the position of the maximum, I_{max}, is shown as a function of time, t, in the double and semilogarithmic plot in Figure 8. Two kinds of time evolution are observed in this system depending on the weight fraction of gelatin. The time evolution of the scattered light intensity in the $\varphi_g = 0.05$ system in the time region around 10 s seems to be expressed by a power law relationship $I_{max} \propto t^\beta$ with $\beta = 1.3$, as shown in the double logarithmic plot in Figure 8. On the other hand, I_{max} increases with an exponential-like growth in the $\varphi_g = 0.1$ and $\varphi_g = 0.2$ systems, as shown in the semilogarithmic plot of Figure 8. Since the time interval of the measurement is limited, we do not strongly claim the above results. It is, however, worth noting here that the time evolution in the spinodal decomposition process is divided into an early and a late stage, and the scattering light intensity increases as a function of time with an exponential-like growth in the very early stage of decomposition. The time interval of the measurements in the $\varphi_g = 0.1$ and $\varphi_g = 0.2$ systems may correspond to the very early stage of the spinodal decomposition. The results may be natural since the gelation reaction accelerates the spinodal decomposition in these two systems. The power law relationship for the $\varphi_g = 0.05$ system may also suggest the presence of the scaling relation, $\beta \geq 3\alpha$ [31]. The detailed discussion of these results requires further precise experimental studies of the system.

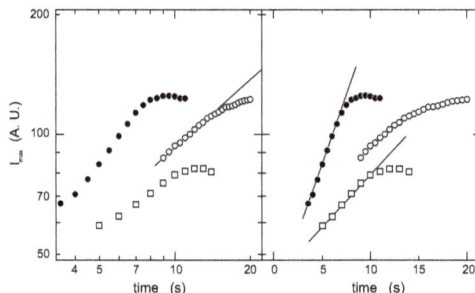

Figure 8. The time evolution of I_{max}. Results are shown for the $\varphi_g = 0.05$ system (open circles), the $\varphi_g = 0.1$ system (closed circles) and the $\varphi_g = 0.2$ system with single quench (open squares). The left figure is the double logarithmic plot of the results. The right figure represents the semi-logarithmic plot of the intensity results. The weight fraction of PEG-1000 is fixed at 0.1. The slope of the line drawn for the $\varphi_g = 0.05$ system in the left figure is about 1.3.

4. Conclusions

The phase behaviors of the ternary system of gelatin, water and PEG oligomer are studied as a function of the weight fraction of gelatin and the weight fraction and molecular weight of PEG oligomer. In addition to the sol-gel transition line, the phase separation line appears in the lower concentration region of the phase diagram when PEG oligomer coexists in the aqueous gelatin solution. The phase separation line is shifted to the higher temperature region upon increasing the concentration and/or the molecular weight of PEG oligomer. The phase behaviors of the system observed under present experimental conditions are well explainable by the site-bond correlated-percolation model of the polymer gelation. Since the phase separation line and the sol-gel transition line cross over in the phase space of $T - \varphi_g$, the phase space is typically divided into four regions. The regions thus appearing in the phase space correspond to the definite phases: one-phase sol, one-phase gel, two-phase sol and two-phase gel.

The transitions between these phases in a system with PEG-1000 at a weight fraction of 0.1 are studied by the light-scattering method. The light-scattering profile from the quenched sample solution shows a maximum at a certain magnitude of the scattering vector, q_{max}. Both the time evolution of the magnitude of q_{max} and the scattering intensity at the maximum, I_{max}, are analyzed.

The time evolutions of q_{max} and I_{max} for the system of $\varphi_g = 0.05$ show the typical behaviors that correspond to the simple spinodal decomposition of the liquid-liquid phase separation process. The results are consistent with the previous optical microscope observations.

The similar behaviors of q_{max} and I_{max} in the $\varphi_g = 0.05$ system are observed for the $\varphi_g = 0.1$ system. The results suggest that the spinodal decomposition also plays a role in the phase transition of this system. However, q_{max} and I_{max} change much earlier and faster than in the $\varphi_g = 0.05$ system, suggesting the acceleration of the spinodal decomposition. Two reasons are considered for the acceleration of the spinodal decomposition. One is the attractive interaction due to the gelation, and the other is the effects of the viscoelastic phase separation.

In contrast to the above two systems, the time evolution of q_{max} is completely altered from the simple spinodal decomposition in the single quench process of the $\varphi_g = 0.2$ system. The results indicate that the sol-gel transition plays a dominant role in all of the transition behaviors. The stepwise quench experiments are done to separate the effects of the sol-gel transition and the phase separation. The results indicate that the phase separation occurs even in the polymer network of the gel. However, the phase separation is limited to a microscopic scale because of the presence of the polymer network of the gel. The results suggest that the phase separation line, when it appears in the gel phase, should be regarded as the microphase separation line.

5. Materials and Methods

Gelatin (Type-B, alkali-treated gelatin, No. 1040781000, Merck, Darmstadt, Germany, M. W. $\simeq 6.9 \times 10^4$) was obtained from Merck and used as obtained. The average molecular weight of gelatin was determined by gel permeation chromatography.

The oligomers of PEG, namely diethylene glycol (M. W. = 106, PEG-100), triethylene glycol (M. W. = 150, PEG-150) and tetraethylene glycol (M. W. = 194, PEG-200) were obtained from Wako Pure Chemical Industry (Osaka, Japan). The polymers of PEG at M. W. $\sim$ 400 (PEG-400) and PEG at M. W. $\sim$ 1000 (PEG-1000), were obtained from Nichiyu Corporation (Tokyo, Japan) and Sanyo Chemical Industry (Kyoto, Japan), respectively.These oligomers and polymers are also used as obtained.

Firstly, water and ethylene glycols were mixed at the desired ratio to make the mixed solvent of gelatin. After ethylene glycol is dissolved completely into water, the calculated amount of gelatin was added into the mixture of water and ethylene glycol. The solutions thus made are heated up to 60 °C to dissolve the gelatin. The obtained solution was transferred into a temperature-controlled bath at a temperature of 60 °C, and then, the temperature is lowered at a rate of 10 °C/h. The solution became opaque near the phase separation temperature. The opaqueness of the solution near the phase separation temperature is caused by the structure change and the formation of the phase-separated droplets by the aggregation of polymers. It has been clarified that the cloud point temperature of the polymer solution practically coincides with the phase separation temperature within an accuracy of a few millidegrees [33,34]. Therefore, the cloud point temperature, which is determined by visual inspection, is assigned as the phase separation temperature in this study. The accuracy of the phase separation temperature thus determined was already confirmed by comparing the data with the light-scattering measurement and the turbidity measurement in previous studies [11,12]. The turbidity of the solution was also measured as a function of the temperature if necessary for the cross-check of the phase separation temperature. After the determination of the phase separation temperature, the solution is kept at 15 °C for 12 h to ensure the gelation. The gelation temperature was determined in this heating process by the falling ball method using a Teflon ball with a wait of 16 mg. Firstly, the Teflon ball was placed on the surface of the gel. The heating rate was chosen as 5 °C/h. The gelation temperature was determined as that at which the ball on the surface of the gel falls into the solution. In this method, the applied stress due to the ball is balanced with the elasticity of the polymer network of the gel. However, the elasticity due to the polymer network diminishes when the system attains the sol-gel transition temperature. The ball falls into the gel when the stress due to the ball overcomes the elasticity of the polymer network. Therefore, the gelation temperature that is determined by the falling ball method corresponds to the fracture temperature of the polymer network of the gel. Although the elastic modulus is not measured, the falling ball method is one of the rheological methods. The weight of the ball should be as small as possible to determine the correct gelation temperature.

It is well known that gelatin is easily degraded by the autocatalysis reaction in the solution state. Thus, it is not favorable to expose gelatin solution to a higher temperature. The cooling rate to determine the phase separation temperature was 10 °C/h, taking into account that the degradation of gelatin should be avoided. Conversely, the slower heating rate was chosen for the exact determination of the gelation temperature by the falling ball method. The phase diagram obtained here, therefore, did not represent an equilibrium state since the cooling rate was rather high. It was, however, sufficient for the present purpose because the phase separation in the liquid state occurs rather quickly. In addition, the thermal hysteresis is not significant in the case of gelatin gel. Further details were given in the previous report [12].

The small angle laser light-scattering measurements were taken using a homemade apparatus. The details of the apparatus are given in [35]. Our apparatus consists of a He-Ne laser (JDSU Uniphase, Milpitas, CA, USA; 8 mW, λ = 632.8 nm) and the one-dimensional detector (Hamamatsu, Japan; S3901-512Q; 512 channels, 50 µm width/channel). The distance between the sample and the detector was changed from about 2 to 30 cm by a one-dimensional translator. Multiple scattering from the sample virtually dominates in the strongly-opalescent region. Such effects are crucial in the detailed discussion near the critical point. In the present system, however, the strong opalescence becomes dominant only in the late stage of the spinodal decomposition. For the present system, a suitable way of minimizing the effects of the multiple scattering is to suppress it by shortening the optical path length in the cell and to work at a larger wavelength because the scattering intensity is proportional to λ^{-4}. The sample gel was, therefore, prepared in the optical cell of a thickness from 1 to 10 mm, depending on the opacity of the gel. The scattering data were obtained by a computer system. The scattering profiles from the gels thus obtained are expressed as a function of the scattering vector, $q = (4\pi n/\lambda)\sin(\theta/2)$, where n, θ and λ are the refractive index, the scattering angle and the wave length of the incident light [36].

Acknowledgments: The authors thank the Ministry of Education, Culture and Sport, Japan, for the supporting this work with a grant-in-aid for for scientific research (KAKENHI, No. 17H06227).

Author Contributions: Miho Yanagisawa and Masayuki Tokita conceived of and designed the experiments. Yutaro Yamashita performed the experiments. Yutaro Yamashita and Masayuki Tokita analyzed the data. Yutaro Yamashita, Miho Yanagisawa and Masayuki Tokita wrote the paper.

Conflicts of Interest: The authors declare no conflict of interest.

Gels **2018**, *4*, 26

Abbreviations

The following abbreviations are used in this manuscript:

PEG	poly(ethylene glycol)
PEG-100	diethylene glycol
PEG-150	triethylene glycol
PEG-200	tetraethylene glycol
PEG-400	PEG at M. W. ~ 400
PEG-1000	PEG at M. W. ~ 1000
q_{max}	magnitude of the scattering vector at the position of the maximum in the scattering profile
I_{max}	scattering light intensity at the position of the maximum in the scattering profile

References

1. Papon, P.; Lebrond, J.; Meijer, P.H.E. *The Physics of Phase Transition, Concepts and Applications*; Chapter 6; Springer: Berlin/Heidelberg, Germany; New York, NY, USA, 2002; ISBN 3-540-43236-1.
2. Djabourov, M.; Lebrond, J.; Papon, P. Gelation of aqueous gelatin solutions. I. Structural investigation. *J. Phys.* **1988**, *49*, 319–332, doi:10.1051/jphys:01988004902031900.
3. Djabourov, M.; Lebrond, J.; Papon, P. Gelation of aqueous gelatin solutions. II. Rheology of the sol-gel transition. *J. Phys.* **1988**, *49*, 333–343, doi:10.1051/jphys:01988004902033300.
4. Malcom, G.N.; Rowlinson, J.S. The thermodynamic properties of aqueous solutions of polyethylene glycol, polypropylene glycol and dioxane. *Trans. Faraday Soc.* **1957**, *53*, 921–931, doi:10.1039/tf9575300921.
5. Tanaka, T.; Swislow, G.; Ohmine, I. Phase separation and gelation in gelatin gels. *Phys. Rev. Lett.* **1979**, *42*, 1556–1559, doi:10.1103/PhysRevLett.42.1556.
6. Nezu, T.; Maeda, H. Phase separation coupled with gelation in polyethylene glycol-gelatin-aqueous buffer system. *Bull. Chem. Soc. Jpn.* **1991**, *64*, 1618–1622, doi:10.1246/bcsj.64.1618.
7. Jizomoto, H. Phase separation induced in gelatin-base coacervation systems by addition of water soluble nonionic polymers I: Microencapsulation. *J. Pharm. Sci.* **1984**, *73*, 879–882, doi:10.1002/jps.2600740420.
8. Jizomoto, H. Phase separation induced in gelatin-base coacervation systems by addition of water soluble nonionic polymers II: Effect of molecular weight. *J. Pharm. Sci.* **1985**, *74*, 469–472, doi:10.1002/jps.2600730705.
9. Morita, T.; Horikiri, Y.; Suzuki, T.; Yoshino, H. Preparation of gelatin microparticles by co-lyophilization with poly(ethylene glycol): Characterization and application to entrapment into biodegradable microspheres. *Int. J. Pharm.* **2001**, *219*, 127–137, doi:10.1016/S0378-5173(01)00642-1.
10. Rahman, M.A.; Khan, M.A.; Tareq, S.M. Preparation and characterization of polyethylene oxide(PEO)/gelatin blend for biomedical application: Effect of gamma radiation. *J. Appl. Polym. Sci.* **2010**, *117*, 2075–2082, doi:10.1002/app.32034.
11. Yanagisawa, M.; Yamashita, Y.; Mukai, S.; Annaka, M.; Tokita, M. Phase separation in binary polymer solution: Gelatin/Poly(ethylene glycol) system. *J. Mol. Liq.* **2014**, *200*, 2–6, doi:10.1016/j.molliq.2013.12.035.
12. Yamashita, Y.; Yanagisawa, M.; Tokita, M. Sol-gel transition and phase separation in ternary system of gelatin-water-PEG oligomer. *J. Mol. Liq.* **2014**, *200*, 47–51, doi:10.1016/j.molliq.2014.03.016.
13. Flory, P.J. *Principles of Polymer Chemistry*; Cornell University Press: London, UK; Ithaca, NY, USA, 1953; ISBN 0-8014-0134-8.
14. Chan, J.W.; Hilliard, J.E. Free energy of a nonuniform system. 1. Interfacial free energy. *J. Chem. Phys.* **1958**, *28*, 258–267, doi:10.1063/1.1744102.
15. Chan, J.W. Phase separation by spinodal decomposition in isotropic system. *J. Chem. Phys.* **1965**, *42*, 93–99, doi:10.1063/1.1695731.
16. Cook, H.E. Brownian motion in spinodal decomposition. *Acta Metall.* **1970**, *18*, 297–306, doi:10.1016/0001-6160(70)90144-6.
17. Hamano, K.; Tachikawa, M.; Kenmochi, Y.; Kuwahara, N. The early stage of phase-separation for the system polydimethylsiloxane diethyl carbonate. *Phys. Lett. A* **1982**, *90*, 425–428, doi:10.1016/0375-9601(82)90800-3.
18. Morita, T.; Narita, T.; Mukai, S.; Yanagisawa, M.; Tokita, M. Phase behaviors of agarose gel. *AIP Adv.* **2013**, *3*, 042128, doi:10.1063/1.4802968.

19. Coniglio, A.; Stanley, H.E.; Klein, W. Site-bond correlated-percolation problem-statistical mechanical model of polymer gelation. *Phys. Rev. Lett.* **1979**, *42*, 518–522, doi:10.1103/PhysRevLett.42.518.
20. Coniglio, A.; Stanley, H.E.; Klein, W. Solvent effects on polymer gel-a statistical-mechanical model. *Phys. Rev. B* **1982**, *25*, 6805–6821, doi:10.1103/PhysRevB.25.6805.
21. Lifshits, I.M.; Slyozov, V.V. The kinetics of precipitation from supersaturated solid solutions. *J. Phys. Chem. Solid* **1961**, *19*, 35–50, doi:10.1016/0022.3697(61)90054-3.
22. Onuki, A. *Phase Transition Dynamics*; Cambridge University Press: Cambridge, UK, 2002; ISBN 0-521-57293-2.
23. Siggia, E.D. Late stage of spinodal decomposition in binary-mixtures. *Phys. Rev. A* **1979**, *20*, 595–605, doi:10.1103/PhysRevA.20.595.
24. Takeno, H.; Hashimoto, T. Crossover of domain-growth behavior from percolation to cluster regime in phase separation of an off-critical polymer mixture. *J. Chem. Phys.* **1997**, *107*, 1634–1644, doi:10.1063/1.474515.
25. Tanaka, H. Viscoelastic phase separation. *J. Phys. Condens Matter* **2000**, *12*, R207–R264, doi:10.1088/0953-8984/12/15/201.
26. Tanaka, H.; Araki, T. Viscoelastic phase separation in soft matter: Numerical-simulation study on its physical mechanism. *Chem. Eng. Sci.* **2006**, *61*, 2108–2141, doi:10.1016/j.ces.2004.02.025.
27. Shi, W.; Liu, W.; Yang, J.; He, Z.; Han, C.C. Hierarchical coarsening in the late stage of viscoelastic phase separation. *Soft Matter* **2014**, *10*, 2649–2655, doi:10.1039/c3sm52713a.
28. Butler, M.F.; Heppenstall-Butler, M. Phase separation in gelatin/dextran and gelatin/maltodextrin mixtures. *Hood Hydrocoll.* **2003**, *17*, 815–830, doi:10.1016/S0268-005X(03)00103-6.
29. Pathak, J.; Rawat, K.; Bohider, H.B. Charge heterogeneity induced binding and phase stability in β-lacto-globulin-gelatin B gels and coacervates at their comon pI. *RSC Adv.* **2015**, *5*, 67066–67076, doi:10.1039/C5RA07195J.
30. Jin, W.; Xu, W.; Ge, H.; Li, J.; Li, B. Coupling process of phase separation and gelation in konjac glucomannan and gelatin system. *Hood Hydrocoll.* **2015**, *51*, 188–192, doi:10.1016/j.foodhyd.2015.05.020.
31. Hashimoto, T.; Takenaka, M.; Izumitani, T. Spontaneous pinning of domain growth during spinodal decomposition of off-critical polymer mixtures. *J. Chem. Phys.* **1992**, *97*, 679–689, doi:10.1063/1.463562.
32. De Gennes, P.G. *Scaling Concepts in Polymer Physics*; Cornell University Press: London, UK; Ithaca, NY, USA, 1979; ISBN 0-8014-1203-X.
33. Kuwahara, N.; Fenby, D.V.; Tamsky, M.; Chu, B. Intensity and line width studies of the system polystyrene-cyclohexane in the critical region. *J. Chem. Phys.* **1971**, *55*, 1140–1148, doi:10.1063/1.1676198.
34. Kojima, J.; Kuwahara, N.; Kaneko, M. Light scattering and pseudospinodal curve of the system polystyrene-cyclohexane in the critical region. *J. Chem. Phys.* **1975**, *63*, 333–337, doi:10.1063/1.431103.
35. Dobashi, T.; Narita, T.; Makino, K.; Mogi, T.; Ohshima, H.; Takenaka, M.; Chu, B. Light scattering of a single microcapsule with a hydrogel membrane. *Langmuir* **1998**, *14*, 745–749, doi:10.1021/la9707641.
36. Hamano, K.; Kuwahara, N.; Nakata, M.; Kaneko, M. Osmotic compressibility and correlation range of the system polystyrene-dimethyl malonate very near its critical point. *Phys. Lett.* **1977**, *63A*, 121–124, doi:10.1016/0375-9601(77)90221-3.

MDPI
St. Alban-Anlage 66
4052 Basel
Switzerland
Tel. +41 61 683 77 34
Fax +41 61 302 89 18
www.mdpi.com

Gels Editorial Office
E-mail: gels@mdpi.com
www.mdpi.com/journal/gels

www.ingramcontent.com/pod-product-compliance
Lightning Source LLC
Chambersburg PA
CBHW051858210326

41597CB00033B/5938